· FROM QUANTUM TO COSMOS ·

Neil Turok is one of the world's leading theoretical physicists and a renowned educational innovator. Director of the Perimeter Institute for Theoretical Physics, he was formerly a professor of physics at Princeton and held a Chair of Mathematical Physics at Cambridge. As well as developing fundamental theories of the cosmos, he has led efforts to determine their predictions and to test them through observation. With Stephen Hawking, he developed the Hawking–Turok instanton solutions describing the birth of inflationary universes. With Paul Steinhardt, he developed a cyclic universe scenario, described in their critically acclaimed book *Endless Universe: Beyond the Big Bang—Rewriting Cosmic History.* In 1992, Turok was awarded the James Clerk Maxwell medal of the UK Institute of Physics and, in 2008, a prestigious TED Prize. Born in South Africa, Turok founded the African Institute for Mathematical Sciences (AIMS), a pan-African network of centres for education and research. This work has been recognized by awards from the World Summit on Innovation and Entrepreneurship (WSIE) and the World Innovation Summit on Education (WISE). Turok lives in Waterloo, Ontario, Canada.

FROM QUANTUM TO COSMOS

The Universe Within

NEIL TUROK

faber and faber

First published in Canada in 2012 by
House of Anansi Press Inc
First published in the UK in 2013
by Faber and Faber Limited
Bloomsbury House
74–77 Great Russell Street
London WC1B 3DA

Printed and bound by CPI Group (UK) Ltd, Croydon, CR0 4YY

A CIP record for this book
is available from the British Library

ISBN 978–0–571–29947–8

FSC
www.fsc.org
MIX
Paper from
responsible sources
FSC® C101712

2 4 6 8 10 9 7 5 3 1

To my parents

CONTENTS

AUTHOR'S NOTE

IN THIS BOOK, I try to connect our progress towards discovering the physical basis of reality with our own character as human beings.

This is not an academic text. I describe some of physics' biggest ideas and how they were discovered, but I make no serious attempt to provide a balanced history or to properly apportion credit. Instead, I use my personal experience as a common thread, along with accounts of people, times, and places that seem special to me. The personalities are interesting, but I use them mainly as illustrations of what is possible and of how much more capable we are than we realize. I am not a philosopher, historian, or an art or literary critic, but I draw on each of these subjects to illustrate the circumstances and the consequences of our deepening knowledge. This is a vast subject, and I apologize for my limited perspective and for my many arbitrary choices.

My goal is to celebrate our ability to understand the universe, to recognize it as something that can draw us together, and to contemplate what it might mean for our future.

I have benefitted from the insights and mentorship of wonderful colleagues, too numerous to mention. I have been equally inspired by many non-scientists, people who through their lives exemplify what it means to be human. Our science and our humanity are two sides of the same coin. Together, they are the means for us to live up to the opportunity of our existence.

ONE

MAGIC THAT WORKS

"Happy is the man who can recognize in the work of
today a connected portion of the work of life, and an
embodiment of the work of Eternity."
— James Clerk Maxwell[1]

WHEN I WAS THREE years old, my father was jailed for
resisting the apartheid regime in South Africa. Shortly
afterward, my mother was also jailed, for six months.
During that time, I stayed with my grandmother, who
was a Christian Scientist. My parents weren't religious,
so this was a whole new world to me. I enjoyed the sing-
ing, and especially the Bible: I loved the idea of a book
that held the answer to everything. But I didn't want a
big Bible; I wanted a *little* Bible that I could carry around
in my pocket.

So I campaigned endlessly for my grandmother to
buy me the smallest possible Bible. When she finally did,

I took it everywhere. I couldn't read yet, but that didn't matter to me. What I most wanted, even at that early age, was to capture and hold the truth, with the certainty and love that it brings.

My father was charged with sabotage and was fortunate to be released after only three and a half years. Others who had been tried on lesser charges were given life terms. On my father's release from prison, he was held under house arrest but escaped and fled north to East Africa. We followed him there and lived in Tanzania for several years before moving to London, England. There we joined a small community of exiles trying to survive in unfamiliar, damp, and gloomy surroundings. Nevertheless, my parents always held firm to their ideas. "One day," they told my brothers and me, "there will be a great change, and South Africa will be free."

It was hard for us to believe them. Throughout the 1970s and 1980s, as I was growing up in England, going to high school and then university, the situation back home seemed hopeless. The apartheid regime was popular with the all-white electorate, and it had powerful allies overseas. South Africa even developed and tested nuclear weapons. The tiny handful of organized dissidents were easily captured and imprisoned. Protests by school students in Soweto were ruthlessly crushed, and the police state took an iron grip.

But then, quite suddenly, everything changed.

The apartheid system was founded on a profoundly wrong premise — that black people are inferior to white people — and this brought its demise. Within the country, the aspirations of the black majority could no longer be contained. External protests also gathered impact as more and more countries boycotted South Africa. In 1993, with Nelson Mandela's negotiated release from prison, the mood turned. The white minority accepted that apartheid was no longer sustainable, and that the future would have to involve universal suffrage and greater opportunities for all. The change in South Africa was wrought by a simple but undeniable idea: justice — the principle of fairness, equity, and human rights that protects us all. Justice is a cause shared across races, cultures, and religions; it is powerful enough to win many people's lifetime commitment and, for some, commitment of their lives. If you had to point to the driver of change in South Africa, it would be this one simple notion that prevailed over all the privilege, wealth, and weaponry that the apartheid regime amassed.

My parents were right. A good idea can change the world.

TODAY, WE LIVE IN a worried world that seems short of good ideas. We are confronted by challenges that can feel overwhelming: financial instability, overconsumption and pollution, energy and resource shortages, climate change, and growing inequality. All of these problems were created by humans, and they are all solvable. Yet we seem to be locked in a culture of short-term thinking, of the quick fix and the fast buck. Whereas what each of these problems really needs for its solution is consistent, principled, far-sighted actions extending over many years.

We're reaching the limits of existing technologies and natural resources. We are in danger of losing our sense of optimism. Can we find smarter ways to manage our planet? Can we make the discoveries that will open up a bright future? Who are we, after all? Are we just the product of a process of random mutation and natural selection, now reaching its terminus? Or are we potentially the initiators of a new evolutionary stage, in which life may rise to a whole new level?

In these chapters, I want to talk about our ability to make sense of reality and to conceive of the universe within our minds. This ability has been a continuous source of powerful ideas, describing everything from the tiniest subatomic particle up to the entire visible cosmos. It has spawned every modern technology, from cellphones to satellites. It is by far and away our most precious possession, and yet it is also completely free to share. If history is anything to go by, the Universe Within us will be the key to our future.

It is not accidental that revolutions take place when they do. The greatest advances have occurred as a result of growing contradictions in our picture of reality that could not be resolved by any small change. Instead, it was necessary to step back, to look at the bigger picture and find a different way of describing the world and understanding its potential. Every time this happened, a whole new paradigm emerged, taking us forward to frontiers we had never previously imagined. Physics has changed the world, and human society, again and again.

The human mind holds these ideas in the balance: how we live together, who we are, and how we place ourselves within reality. Our conceptions greatly exceed any immediate need. It is almost as if the evolutionary process has an anticipatory element to it. Why did we evolve the capacity to understand things so remote from our experience, when they are seemingly useless to our survival? And where will these abilities take us in the future?

How did we first imagine the Higgs boson, and build a microscope — the Large Hadron Collider, capable of resolving distances a billionth the size of an atom — to find it? How did we discover the laws governing the cosmos, and how did we build satellites and telescopes that can see ten trillion times farther than the edge of the solar system, to confirm those laws in detail? I believe society can draw great optimism from physics' phenomenal success. Likewise, physics can and should draw a greater sense of purpose from understanding its own origins, history, and connections to the interests of society.

What is coming is likely to be even more significant than any past transformation. We have already seen how mobile communications and the World Wide Web are opening up global society, providing information and education on a scale vastly larger than ever before. And this is only the beginning of how our new technologies will change us. So far, our scientific progress has been founded on, but also limited by, our own physical nature. We are only able to comprehend the world in a classical picture. This has been an essential stepping stone in our development, but one that we need to move beyond. As our technological capabilities grow, they will drastically extend our abilities, our experience of the world, and, in time, who we are.

The internet is only a harbinger. Quantum technologies may change entirely the way in which we process information. In time, they may do much more, allowing us to gain a heightened awareness of reality and of the ways the physical world works. As the depth of our knowledge grows, our representations of the universe will achieve much higher fidelity. Our new knowledge will enable technologies that will vastly supercede current limits. They may change our very nature and bring us closer to realizing the full potential of our existence.

As we look ahead our goal should be to experience, to understand, and to be a part of the universe's development. We are not merely its accidental byproducts; we are the leading edge of its evolution. Our ability to explain the world is fundamental to who we are, and to our future. Science and society's mission should be one and the same.

Engraved on Karl Marx's tombstone are these famous words: "The philosophers have only interpreted the world, in various ways. The point, however, is to change it." Riffing on a quote attributed to Gandhi, I would say, "The point, however, is to *be* the change."

I HAVE BEEN FORTUNATE to spend some of my life in Africa, the cradle of humanity. One of my most memorable experiences was visiting the Ngorongoro Crater, the Serengeti, and the Olduvai Gorge, which early human ancestors inhabited nearly two million years ago. There is an abundance of wild animals — lions, hyenas, elephants, water buffalo. Even the baboons are dangerous: a large male weighs nearly a hundred pounds and has enormous incisors. Nevertheless, they are all afraid of humans. If you are camping there and a big baboon tries to steal your food, all you have to do is raise your arm with a stone in your hand, and it will scurry away.

As puny as human beings are, our ancestors acquired dominance over the rest of the animal kingdom. With their new modes of behaviour, standing up and throwing stones, using tools, making fires, and building settlements, they outsmarted and out-psyched all the other creatures. I have seen elephants and water buffalo move away at merely the scent of a lone Maasai approaching, strolling unconcernedly through the bush with his hand-held spear, as if he were the king of it all. Our mastery of nature began with our ancestors in Africa, and they deserve our utmost respect.

From the invention of tools and then agriculture, the next great leap forward in technology may have been the development of mathematics: counting, geometry, and other ways of understanding regularities in the world around us. Many of the oldest mathematical artifacts are African. The oldest is a baboon's leg bone, from a cave in Swaziland, dated to 35,000 B.C. It has twenty-nine notches on it, perhaps marking the days of a lunar cycle. The second oldest is another baboon's leg bone found in the eastern Congo and dating from around 20,000 B.C. It is covered in marks grouped in a manner suggestive of simple arithmetic. The oldest known astronomical observatory is a stone circle at Nabta Playa, in southern Egypt near the border with Sudan, built around 4000 B.C. Then, of course, there are the great pyramids built in Egypt, from around 3000 B.C. onwards. Mathematics allowed people to reliably model the world, to make plans, and to predict outcomes.

As far as we know (no written records survive), the idea that mathematics could reveal powerful truths about the universe originated with Pythagoras and his followers in ancient Greece in the sixth century B.C. They invented the word "mathematics" (the Pythagoreans were called the *mathematici*) and the notion of a mathematical "proof": a set of logical arguments so compelling as to make the result unquestionable. The Pythagorean theorem — that the area of a square drawn on the long side of a right-angled triangle equals the combined area of the squares drawn on the two shorter sides — is the most famous such proof. (However, the fact it proves was known much earlier: for example, it is referred to on tablets used for surveyors' calculations dating from around 1800 B.C. in ancient Babylon, near modern Baghdad.)

The Pythagoreans formed a religious cult, based near Croton, in southern Italy, with a focus on mathematics' mystical power. One of their accomplishments was to understand the mathematical nature of music. Dividing a plucked string into halves produced an octave, into thirds a fifth above that octave, and into quarters another fourth above that. If mathematics could so neatly account for musical harmonies, they reasoned, it might explain many other aspects of order in the universe. Building on earlier ideas of Anaximander, who some consider to have been the first scientist and was, perhaps, Pythagoras's teacher, the Pythagoreans attempted to "construct the whole heavens out of numbers."[2] This insight, two millennia before Newton, was to become the foundation for all of physics.

The Pythagoreans apparently gave good advice to Croton's rulers — for example, on introducing a constitution — which helped the town's economy prosper. But they were also perceived as elitist and obsessively secretive. In the words of one historian, "Their assumption of superiority and esoteric knowledge must at times have been hard to bear."[3] This probably contributed to the Pythagoreans' tragic downfall, in which, according to some versions, Pythagoras was killed. The Pythagoreans' untimely demise was an early signal of the dangers of the separation of scientists from everyday society.

The division has reappeared, again and again. For example, in medieval Europe, the university bachelor's curriculum was dominated by Latin, logic, and rhetoric (the trivium), the skills which were needed for diplomacy, government, and public presentation. Those who continued to their master's would take arithmetic, music, geometry, and astronomy (the quadrivium). The separation between non-scientists and scientists was inevitable, as human knowledge expanded and expertise became more and more specialized. It led to a division between the sciences and the arts and humanities, which the English physicist and author C. P. Snow famously referred to as the "Two Cultures." This seems to me unfortunate. Isn't science also an art? And shouldn't scientists also have humanity?

I recall being upset as a young scientist by the words of one of my heroes, the great U.S. physicist Richard Feynman, who recounted how he overcame his worries about working on the nuclear bomb: "[The Hungarian-American mathematician John] von Neumann gave me an interesting idea: that you don't have to be responsible for the world that you're in. So I have developed a very powerful sense of social irresponsibility as a result of von Neumann's advice. It's made me a very happy man ever since. But it was von Neumann who put the seed in that grew into my *active* irresponsibility!"[4] At the time, Feynman's cop-out seemed incompatible with what I knew of his persona. His humanity shone through in his writing, his teaching, and all his interactions. Only later, I realized he was in denial. Feynman loved his physics; he just couldn't face thinking about the far more difficult questions of the uses to which it might be put.

THE DISCONNECTION BETWEEN SCIENCE and society is harmful, especially when you consider that science is, in general, open-minded, tolerant, and democratic. In its opposition to dogma and its willingness to live with uncertainty, science is in many ways a model for society. Many scientists are energized by the sense that their work is of wider interest and might contribute to progress. Back in the eighteenth century, the Scottish philosopher David Hume wrote these wise words: "It seems, then, as nature has pointed out a mixed kind of life as most suitable to the human race...Indulge your passion for science, says she, but let your science be human, and such as may have a direct reference to action and to society."[5] Equally, he argued, society in its aesthetic and moral concerns can benefit from science: "Accuracy is in every case advantageous to beauty, and just reasoning to gentle sentiment."[6]

Hume had entered Edinburgh University as a lad of twelve — starting so young was not uncommon at the time — during the period known as the Scottish Enlightenment. His independence of mind is nicely illustrated in a letter he wrote at the end of his time at Edinburgh: "There is nothing to be learnt from a Professor, which is not to be met with in Books."[7] Nevertheless, it was at university that he discovered his passion for philosophy. He spent the eight years following his graduation writing his philosophical masterpiece, *A Treatise of Human Nature*, the first volume of which would later appear as *An Enquiry Concerning Human Understanding*.

Hume's *Enquiry* reads, even today, as a breath of fresh air. His modesty, his originality, his accessible style are models of the art of gentle persuasion. His powers of reason worked wonders as he calmly overturned two millennia of doctrinaire thinking.

Hume's revolutionary views, though far-reaching, were based on the simple suggestion that our existence, our feelings, and our experience are the foundation for all our ideas. Imagination is powerful, but it is no substitute for our natural impressions and instincts: "The most lively thought is still inferior to the dullest sensation,"[8] and again, "It is impossible for us to *think* of any thing, which we have not antecedently *felt*, either by our external or internal senses." Even mathematical abstractions like number or shape are, Hume argued, ultimately based upon our experience of interacting with natural phenomena.[9]

Hume believed our perceptions and feelings — our external and internal experiences — to be the foundation for our knowledge. It was a profoundly democratic idea, that knowledge is based on capacities which everyone shares. While recognizing the power of mathematics, Hume warned against reasoning too far removed from the real world: "If we reason *a priori*, anything may appear able to produce anything. The falling of a pebble may, for aught we know, extinguish the sun; or the wish of a man control the planets in their orbits. It is only experience, which teaches us the nature of cause and effect, and enables us to infer the existence of one object from that of another."[10] In his constant emphasis on experience, Hume helped to bring science back to earth, to connect it to our humanity, to who we are and what we can do.

Hume's skepticism and frankness brought him into conflict with the Church. Hume's *Dialogues Concerning Natural Religion* (echoing Galileo's *Dialogue Concerning the Two Chief World Systems*) is framed as a debate between three protagonists in an ancient Greek setting, *Dialogues* addresses the validity of beliefs, such as the existence of a creator, the immortality of the soul, and the moral benefits of religion. It does so in a subtle and respectful way, encouraging open discourse without belittling the protagonists. Nevertheless, even as they recognized the book as a landmark, Hume's friends persuaded him that it would be dangerous to publish. The book only appeared in print three years after his death, anonymously and with no publisher credited.

Hume took a unified approach to natural and moral philosophy — which he called "the science of man." He expressed a balanced view of the advantages and limitations of both: "The chief obstacle, therefore, to our improvement in the moral or metaphysical sciences is the obscurity of the ideas, and the ambiguity of the terms. The principal difficulty in the mathematics is the length of inferences and compass of thought, requisite to the forming of any conclusion. And perhaps, our progress in natural philosophy is chiefly retarded by the want of proper experiments and phaenomena, which are often discovered by chance, and cannot always be found, when requisite, even by the most diligent and prudent enquiry."[11] In this last point, he had great foresight. In the nineteenth century, experiments and observations drove an age of discovery. Even in the twentieth century, Einstein was influenced by Hume and expressed his core beliefs in the same language.[12]

What Hume put forward continues to resonate today. Our ability to do science is rooted in our relationship with the universe, our nature as living beings. Our feelings and instincts are far more profound than our ideas. Our ideas allow us to imagine many things, but they can be unreliable, misguided, or misleading. It is the real world that keeps us honest.

Science is about discovering things: about the universe and about ourselves. We are looking for answers, for explanations that will open new doors. What is the meaning of life in the universe, or the purpose of our existence? Scientists typically refrain from any discussion of these notions, saying they are beyond science's realm. But to me, such questions are profoundly important. Why do we decide to do the things we do? Are we, as some scientists would say, merely biological machines, driven by the need to replicate our selfish genes? If we can, as I believe, be much more than this, from where can we draw our wisdom?

Hume's philosophy of knowledge was closely connected to his notions of ethics and of society. Our strength as scientists rests on our character and honesty as human beings, the same traits that make us good citizens. And all of these capabilities arise from our connection with the universe.

· · ·

AS A CHILD, I spent many hours watching ants, amazed at how these tiny creatures determinedly followed paths away from or back to their nests, and wondering how they coped with unexpected changes, like a stick across their path, being soaked in rain, or being blown off course. Like us, they must be constantly extracting the essential information they need from their surroundings, updating their mental models of the world, weighing their options, and taking decisions.

Our brains seem to work this way. We each have an internal model of the world, which we are constantly comparing against our perceptions. This internal model is a selective representation designed to capture reality's most essential elements, the ones that are most important for us, and predict their behaviour. In receiving data from our senses, what we notice are the surprises — the discrepancies between our experiences and the predictions of our internal model, which force us to correct it. Science is the extension of this instinctive ability, allowing us to create explanatory knowledge at ever deeper and more far-reaching levels.

Mathematics is one of our most valuable tools, and perhaps *the* most valuable tool, in this reduction of nature down to its key elements. It is founded on mental abstractions like number, shape, and dimension, which are distillations of the properties of objects in the real world. It complements our natural instincts and intuition in magical, unexpected ways. For example, when perspective and shadowing, which are entirely geometrical concepts, were first employed by artists in medieval Italy, paintings suddenly leaped from the flat two-dimensional world of medieval icons to the infinitely richer three-dimensional world of Renaissance art.

Leonardo da Vinci mastered these techniques, combining art and science in equal measure. Most famous for his paintings — some of the finest ever made — he also made a great number of drawings, of imaginary machines and inventions, of plants and animals, and of cadavers dissected illegally to reveal the inner workings of the human body.

Leonardo never published his writings, but he did keep personal notes that survived, although in complete disorder. Written in mirror-image cursive, from right to left, they open with this rejection of authority: "I am fully aware that the fact of my not being a man of letters may cause certain presumptuous persons to think that they may with reason blame me, alleging that I am a man without learning. Foolish folk! ... they do not know that my subjects require for their exposition experience rather that the words of others."[13]

He was not at all against theory, however — on the contrary, he states: "Let no man who is not a Mathematician read the elements of my work."[14] And elsewhere: "The Book of the science of Mechanics must precede the Book of useful inventions."[15] Like the ancient Greeks, he was strongly asserting the power of reason.

As an artist, Leonardo was understandably obsessed with light, perspective, and shadow. In his notebooks he explained how light is received, with the eye at the apex of a "pyramid" (or cone) of converging straight rays. Likewise, he discussed in detail how shadows are produced by the obstruction of light. Many of his mathematical ideas may be traced back to those of Alhazen (Ibn al-Haytham, 965–1040), one of the most famous Islamic scientists, who worked in both Egypt and Iraq at the end of the first millennium and whose *Book of Optics* (*Kitab al-Manazir*), written in 1021, was published in Italy in the fourteenth century.

Leonardo's careful use of geometry and scientific employment of perspective and shadow, as well as his deep appreciation of anatomy, allowed him to create stunning works of art which not only captured the real world but playfully represented imaginary landscapes (as in the background of the *Mona Lisa*) or historical scenes (like *The Last Supper*). To see the effect of these advances, one has only to look at the way art was transformed. Before the Renaissance, paintings were little more than cartoon representations of the world; after it, realistic representations became normal.

Mathematics can take us far beyond our natural instinct for understanding the world. A mathematical model is a representation of reality, which we improve by an iterative process of trial and error, adaptation and refinement. Our models evolve, much as life does, and as they develop they change and are steadily improved. They are never final. As Einstein said, "As far as the laws of mathematics refer to reality, they are not certain: and as far as they are certain, they do not refer to reality."[16] Stated differently, being creatures of limited capability living in a very complex world, the best we can do is to focus on and understand nature's underlying regularities.

From the motion of the planets, to the structure of atoms and molecules, to the expansion of the cosmos, many of the world's most basic properties are accurately predictable from beautifully simple mathematical rules. Italian mathematician Galileo Galilei is reported to have said, "Mathematics is the language with which God wrote the universe."[17] It is an especially powerful language, a set of logical rules that allow no contradiction.

As an example, the circumference of a circle is its diameter times a number called π (pi). π is a peculiar number, first estimated by the Babylonians as about 3, shown by Greek scientist Archimedes (287–212 B.C.) to be between $3\frac{1}{7}$ and $3\frac{10}{71}$, then approximated by a Chinese mathematician, Zu Chongzhi (A.D. 429–500), as $\frac{355}{113}$. But the point is, it doesn't matter which circle you choose, π always comes out the same—3.14159..., with digits that go on and on and never repeat themselves. Well, all right, you say, π is a useful little rule. And handily enough, it turns up again in the volume of a sphere, any sphere of any size, anywhere in the universe — from a basketball to a planet. In physics, it turns up everywhere: in the formula for the period of a pendulum, or the force between two electric charges, or the power in a shockwave. And that's only the beginning.

We do not understand why mathematics works to describe the world, but it does.[18] One of its most remarkable features is that it transcends culture or history or religion. Whether you are Mexican or Nigerian, Catholic or Muslim, speak French or Arabic or Japanese, whether you lived two millennia ago or will live two millennia in the future, a circle is round and two plus two is four.

The reliable, seemingly timeless character of mathematical knowledge has allowed us to build our societies. We count, plan, and draw diagrams. From water and electrical supplies; to architecture, the internet, and road-building; to financial, insurance, and market projections; and even to electronic music, mathematics is the invisible plumbing of modern society. We normally take it for granted, and we don't notice it until the pipes burst. However, mathematical models are only as good as their assumptions. When those assumptions are faulty or corrupted by wishful thinking or greed, as they were in the recent financial crisis, our whole world fails with them.

PHYSICISTS, ON THE OTHER hand, are interested in discovering the basic laws that govern the universe. Theoretical physics is the application of mathematics to the fundamental description of reality. It is the gold standard of mathematical science, and our most powerful internal model of the world.

Again we return to late sixteenth- and early seventeenth-century Renaissance Italy, where Galileo Galilei took the first steps towards founding the field of physics. He realized that mathematics, when used in conjunction with careful experiments and accurate measurements, could provide a powerful description of the real world. Mathematics allows us to form conceptions of the world far beyond our everyday experience, to delve deeply into our models of reality, and to search for contradictions in our descriptions, which often suggest new phenomena. But in the end, the only true test of the correctness or falsity of our ideas is, as Galileo first fully appreciated, experiment and observation.

So, through a combination of logical reasoning, observation, and painstaking experiment, Galileo developed physics as a new, universal discipline. His experiments with soot-blackened balls rolling on inclined planes, and his observations of the moons of Jupiter and the phases of Venus, provided the vital clues to ruling out the ancient Ptolemaic picture, in which the Earth lay at the centre of the universe, and establishing instead a Copernican universe with the sun at the centre of the solar system. That was the first step on the path to a Newtonian universe.

Galileo was a prodigious inventor: of a geometrical compass, a water clock, a new type of thermometer, telescopes, and microscopes, all the instruments that allowed him to accurately observe and measure the world. He risked his life in pursuit of his ideas. His notion of universal mathematical laws of motion, which could be uncovered by reason, was very threatening to religious authority. When his observations supported the Copernican, heliocentric picture of the solar system and directly contradicted the views of the Catholic Church, he was tried by the Inquisition and was forced to recant and then to live under house arrest for the rest of his life. He used the period of his imprisonment to write his final masterpiece, *Two New Sciences*, which laid the ground for Newton's theory of mechanics. These achievements inspired Albert Einstein to call Galileo "the father of modern physics — indeed of modern science."[19]

The combination of mathematical theory and real experience, pioneered by Galileo, drove the development of every modern technology, from electronics to construction engineering, from lasers to space travel. And it opened up the universe to our understanding, from far below the size of an atom right up to the entire visible cosmos. To be sure, there are still great gaps in our knowledge. But when we look at how rapidly and how far physics has come since Galileo, who can say what its future limits are?

· · · ·

MY OWN ATTRACTION TO maths and physics began when I was about seven years old. Upon my father's release from prison in 1966, he realized he was in serious danger of rearrest. So he escaped across the border with Botswana and made his way overland to Kenya. After a considerable delay, my mother, my two older brothers, and I were granted permission to join him, under the condition that we never return to South Africa. However, as a refugee, my father had no passport and could not obtain employment. Neighbouring Tanzania, under President Julius Kambarage Nyerere, was far more strongly committed to supporting the struggle against apartheid. So, after a brief stay in Nairobi, we were granted asylum in Tanzania and moved to Dar es Salaam, the country's largest city.

I was sent to a government school, where I had a wonderful Scottish teacher named Margaret Carnie. She encouraged me to undertake many scientific activities, like making maps of the school, building electric motors, and playing around with equations. She was passionate about teaching, extraordinarily supportive and not at all prescriptive, and she gave me a lot of freedom. Most of all, she believed in me.

When I was ten, we moved to London, England, just in time to see the Apollo 11 lunar landing and watch Neil Armstrong step onto the moon. Who could ever forget the picture of Earth as a gorgeous blue marble floating above the moon's horizon? We were swept up in the moment and filled with optimism for the future.

It was the end of the sixties, and space was suddenly the coolest thing around. It's hard to convey the sense of the excitement, how the space program bound together people from all walks of life and every political opinion.

It symbolized a certain spirit — ambitious and aglow with the crazy idea of using technology to fling a climbing rope up to the cosmos.

Equally as enthralling as the moon landing was the drama of Apollo 13, only one year later. Imagine you're 320,000 kilometres from home, out in the void of deep space, and you hear a loud bang. "Houston, we have a problem…" One of two main oxygen tanks had exploded, leaking precious oxygen into space over the next two hours. The three astronauts crowded into the only lifeboat they had: the little lunar explorer capsule, which had nowhere near the fuel they needed to get back to Earth. The drama was incredible. There were daily bulletins on TV. All over the world, people were biting their nails. How could the astronauts possibly survive?

NASA's engineers came up with a fantastic solution. They used the moon's gravity to pull them towards it, then slingshot the little pod around its dark side and back to Earth. A few days later, there the astronauts were, their hot little tin can dive-bombing into the Pacific, where they were fished out and then, incredibly, waving to us from the TV, gaunt, unshaven, but alive. Everyone survived. It was pure magic.

The trajectory for this manoeuvre was computed using the equations discovered by the founder of the field we now call theoretical physics, and also one of the most capable mathematicians of all time: Isaac Newton.

Newton, like Galileo, was an outsider. He came from an ordinary background but possessed an extraordinary mind. He was deeply religious but highly secretive about his beliefs. And understandably so, since, for example, he passionately rejected the idea of the Holy Trinity while spending the duration of his scientific career at Trinity College in Cambridge. Newton seems also to have been motivated to a large degree by mysticism — he wrote far more on interpretations of the Bible and on the occult than he ever did on science. The famous economist John Maynard Keynes studied Newton's private papers, a box of which he had acquired at auction, and came to this conclusion: "Newton was not the first of the age of reason. He was the last of the magicians, the last of the Babylonians and Sumerians, the last great mind which looked out on the visible and intellectual world with the same eyes as those who began to build our intellectual inheritance rather less than 10,000 years ago."[20]

Newton spent most of his early scientific years on alchemy, researching transmutation (turning base elements into gold) and trying to find the elixir of life. None of these efforts were successful; he seems to have succeeded only in poisoning himself with mercury. This poisoning may have contributed to a nervous breakdown he is believed to have suffered around the age of fifty-one, after which he largely gave up doing serious science.

Newton's mathematical researches were his magic that worked. He searched for mathematical formulae that would describe the motion of objects on Earth and the planets in space. He found spectacularly simple and successful answers. In the late sixteenth century, a series of very accurate measurements of the motions of celestial bodies were made by the astronomer Tycho Brahe from the world's greatest observatory of the time, Uraniborg in Denmark. Brahe's protégé, German mathematician and astronomer Johannes Kepler, had successfully modelled the data with some ingenious empirical rules. It fell to Newton to develop Galileo's insights into a complete mathematical theory.

BEFORE GALILEO, COPERNICUS HAD pioneered the idea that the Earth was not the centre of the universe. The prevailing wisdom, tracing back to Aristotle and Ptolemy, held that the sun, moon, and planets moved around the Earth carried on a great interlocking system of celestial spheres, which could be carefully arranged to fit the observations. Aristotle claimed that it was just in the Earth's nature not to move. Earthly bodies followed earthly laws, and celestial bodies obeyed celestial laws.

Newton's point of view was quite different: his law of gravitation was the first step on a path towards "unification," a single, neat set of mathematical laws describing all of physical reality. It was the most far-reaching idea, that exactly the same laws should apply everywhere — on Earth, in the solar system, right across the cosmos. Newton's law of gravitation states that the gravitational force of attraction between two objects depends only on their masses and how far apart they are. The more massive the object, the more strongly it attracts and is attracted. The farther apart two objects are, the weaker the force of attraction between them.

In order to work out the consequences of this law of gravity, Newton had to develop a theory of forces and motion. It required a whole new type of mathematics, called "calculus." Calculus is the study of continuous processes, such as the motion of an object whose position is given as a function of time. The velocity measures the rate of change of the object's position, and the acceleration tells you the rate of change of the velocity. Both are calculated over infinitesimally small times, so calculus implicitly rests on a notion of infinitely small quantities. Once he had developed it, Newton's theory had applications well beyond gravity or the solar system. It predicts how *any* collection of objects will move when *any* set of forces acts upon it.

In describing the motion of objects, Newton's starting point was an idealization. How would an object behave if it were released in empty space, with nothing else around it? To be specific, picture a hockey puck floating all alone in an absolute void that stretches to infinity. Let's ignore gravity, or any other forces. What would you expect the puck to do? If it was all alone, and there was nothing nearby to measure its position from, how could you tell if it was moving?

Now imagine a second hockey puck, also floating freely in the void. Picture two tiny people, each of them standing on one of the pucks and seeing the other puck some way off. What do they see? And how will each puck move?

Newton's answer was simple. According to the view

from either puck, the other puck will move in a straight line and at a constant speed, forever. If you imagine more and more pucks, with none more special than any other, then according to every puck's viewpoint, every other puck will move in the same way. This was Newton's first law of motion: in the absence of forces, the velocity of any object remains constant.

Let us come back down to earth, to a perfectly smooth, slippery ice rink. The world's greatest Zamboni has just gone over it. Imagine a puck sliding along the ice in a perfectly straight line. But now you skate alongside it and push it with your stick. Push on its side and the trajectory will curve; push behind it and you can speed it up. Newton's second law describes both effects in one equation: force equals mass times acceleration.

Finally, when you push on anything — the puck, another person, or the side of the rink — it pushes back at you equally hard. This is described by Newton's third law, which says that for every force there is always an equal and opposite force.

Newton's three laws are simple but incredibly powerful. They describe everything known about motion prior to the twentieth century. In combination with his law of gravitation, they explain how the force due to the sun's gravity pulls the planets inwards — just as a string pulls on a whirling stone — and bends the motion of the planets into orbit around it. According to Newton's third law, just as the string pulls in the stone or the sun pulls the Earth around it, the stone pulls the string out and the Earth pulls back on the sun, causing the sun's position to wobble back and forth slightly as the Earth goes around it. The same effect is now used to search for planets in orbit around other stars: the slight wobble in a distant star's position causes a tiny modulation in the colour of the light we receive, which can be detected. More familiar is the effect of the moon's gravitational pull on the water in Earth's oceans, which is responsible for the tides.

Implicit within these laws was the idea, dating back to Galileo, that it is only the *relative* positions and motions of objects that really matter. Galileo pointed out that a person travelling in the hold of a ship, which is sailing steadily along, simply cannot tell from watching anything inside the ship — for example, a fly buzzing around — whether the ship is moving. Today we experience the same thing when we sit in an aircraft moving at 1,000 kilometres per hour and yet everything feels just as if we are at home in our living room.

In our ice-rink world, we can see the same effect. Imagine two pucks that happen to be sliding along the ice exactly parallel to each other and moving at the same speed. From either puck's point of view, the other is not moving. However, from a third puck's point of view, both would be moving in straight lines at the same velocity. In this ice-rink world, all that really matters are the *relative* positions and motions of the objects. Because Newton's laws never mention a velocity, the point of view of any puck moving at any constant velocity is equally valid.

All such observers agree on forces and accelerations, and they would all agree that Newton's laws are valid.

The idea that the same laws of motion apply for any observer moving at a constant velocity was very important. It explained how it can be that we are moving rapidly through space around the sun without feeling any effect. Our orbital speed is huge — around 30 kilometres per second — but, as Galileo realized, it is imperceptible to us because everything around us on the surface of the Earth is travelling right alongside us, with exactly the same enormous velocity. Today we know that the sun is moving, at an even more fantastical speed of 250 kilometres per second, around our galaxy, and that our galaxy, the Milky Way, is moving at a yet greater speed of 600 kilometres per second through the universe. We are actually space travellers, but because Newton's laws do not care about our velocity, we don't feel a thing!

Newton's law of gravity describes with exquisite precision the invisible, inexorable tie that binds the seat of your pants to your chair, holds the Earth and the planets in orbit around the sun, holds the stars in their spherical shape and keeps them in their galaxies. At the same time, it explains how Earth's gravity affects everything from baseballs to satellites. That exactly the same laws should apply in the *un*earthly and hitherto divine realm of the stars as in the imperfect human world around us was a conceptual and indeed a spiritual break with the past. As Stephen Hawking has said, Newton unified the heavens and the Earth.

Newton's laws are as useful as ever. They are still the first rules that every engineer learns. They govern how vehicles move, on Earth or in space. They allow us to build everything from machines and bridges to planes and pipelines — not just by crafting, eyeballing, and adjusting, but by design. Although Newton discovered his laws by thinking about the motion of planets, they enabled the development of a vast number of technologies here on Earth, from bridge building to the steam engine. His notion of force was the key to all of it. It explained how we, through controlling and governing forces, could harness nature to our purposes.

More than three centuries after he published his findings in *Mathematical Principles of Natural Philosophy*, known as the *Principia*, Newton's universal laws of motion and gravitation are still the foundation for much of engineering and architecture. His discoveries underpinned the Industrial Revolution that transformed the organization of human society.

The universe that Newton's laws describe is sometimes called the "classical" or "clockwork" universe. If you know the exact position and velocity of every object at one time, then in principle Newton's laws predict exactly where every object was or will be at any time in the past or future, no matter how remote. This classical universe is completely deterministic, and it is straightforward and intuitive. But as we shall see, in this respect it is utterly misleading. Before we get to that part of the story, we must discuss another outsider who, two hundred years later, would make a discovery even greater than Newton's.

. . .

THE STORY OF THE discovery of the nature of light begins, appropriately enough, with the great flowering of intellectual thought known as the Scottish Enlightenment. At the turn of the eighteenth century, after a dark and brutal period of domination by the monarchy and the Catholic Church, England was preoccupied with building the British Empire in Africa, the Americas, and Asia, giving Scotland the space to establish a unique identity. Scotland emerged with a powerful national spirit, determined to set its own course and to create a model society. Scotland's parliament founded a unique public school system with five hundred schools, which, by the end of the eighteenth century, had made their country more literate and numerate than any other in the world. Four universities were founded — in Glasgow, St. Andrews, Edinburgh, and Aberdeen — and they were far more affordable than Oxford or Cambridge, the only universities in England. The Scottish universities became centres of public education as well as academic study.

Edinburgh became the leading literary centre in Europe and home to luminaries such as David Hume and the political philosopher Adam Smith. According to Arthur Herman, author of *How the Scots Invented the Modern World*, it "was a place where all ideas were created equal, where brains rather than social rank took pride of place, and where serious issues could be debated... Edinburgh was like a giant think tank or artists' colony, except that unlike most modern think tanks, this one was not cut off from everyday life. It was in the thick of it."[21]

Scottish academia likewise followed a distinct course, emphasizing foundational principles and encouraging students to think for themselves, explore, and invent. There was a lively debate, for example, over the meaning of basic concepts in algebra and geometry, and their relation to the real world.[22]

This focus on the fundamentals was remarkably fruitful. As just one instance, English mathematician and Presbyterian minister Reverend Thomas Bayes, whose famous "Bayes theorem" was forgotten for two hundred years but now forms the basis for much of modern data analysis, attended Edinburgh University at the same time as Hume. Fast on the heels of Scotland's academic flowering came the great Scottish engineers, such as James Watt, inventor of the steam engine, and Robert Stevenson, who built the Bell Rock Lighthouse, off the coast of Angus, Scotland.

As the Western world entered the nineteenth century, the Industrial Revolution permeated and remade every aspect of life. The power of steam engines revolutionized the economy. Distances shrank with trains, steamships, and other conveyances; people moved en masse into cities to work in factories that made everything from textiles to pots and pans and that in so doing redefined notions of both work and economic value. A new breed of "natural philosophers" — mainly gentleman hobbyists — set out to understand the world in ways that had never before been possible. The effect of the Scottish Enlightenment was felt at the highest levels of science. Having spawned philosophers, writers, engineers, and inventors, Scotland now produced great mathematicians and physicists. One particular young genius would expose nature's inner workings to a degree that outshone even Newton.

Newton's physics explains a great many things, from the ebb and flow of the tides, caused by the moon's gravitational attraction, to the orbits of planets, the flow of fluids, the trajectories of cannonballs, and the stability of bridges — everything involving motion, forces, and gravity. However, Newtonian physics could never predict or explain the transmission or reception of radio waves, the telephone, electricity, motors, dynamos, or light bulbs. The understanding of all this, and a great deal more, we owe to the experimental work of Michael Faraday, born in 1791, and its theoretical elaboration by James Clerk Maxwell, born four decades later.

One can see the pair, Faraday and Maxwell, as the experimental yin and the theoretical yang of physics. Together, they typify the golden age of Victorian science. The well-born, well-educated Maxwell (he was heir to a small Scottish estate) fits a definite type: a gentleman scientist who, largely freed from the pressures of earning a living, pursued science as an ardent, passionate hobbyist.

James Clerk Maxwell was a bright and curious child, born in southern Scotland. Having the run of his family's estate in Glenlair, he was interested in everything natural and man-made. "What's the go o' that?" he asked, again and again, picking up insects or plants or following the course of a stream or a bell-wire in the house. Joining a private school — the Edinburgh Academy — at age ten, he was known as "Dafty" and bullied, in part for his strange clothes, designed by his father who, though a lawyer by profession, was scientifically minded. By fourteen, with his father's encouragement, Maxwell had become a keen mathematician, preparing a paper describing a new way to draw ovals, which was read to the Royal Society of Edinburgh by a local professor.

The Scottish educational system was particularly strong in mathematics. Rather than learning mathematics by rote as what one professor contemptuously termed a "mechanical knack," students worked through the fundamentals from first principles and axioms. When James Clerk Maxwell found his first great friend, Peter Guthrie Tait, as a schoolkid, they amused themselves by trading "props," or "propositions" — questions they'd make up to try to outwit one another. It became their bond, and decades later, when they were both eminent physicists, Maxwell continued to send his old friend questions that stumped him and whose answers helped him piece together the puzzle of electromagnetism.

Maxwell, Tait, and William Thomson — later Lord Kelvin — who was educated at Glasgow University, formed a Scottish triumvirate, with all three becoming leading physicists of their time. Tait and Thomson co-authored the *Treatise on Natural Philosophy*, the most important physics textbook of the nineteenth century. Tait founded the mathematical theory of knots and Lord Kelvin made major contributions to many fields, including the theory of heat, where his name is now attached to the absolute scale of temperature. Alexander Graham Bell, another great Scottish inventor, followed Maxwell to university in Edinburgh before emigrating to Canada and developing the telephone.

After three years at Edinburgh University, Maxwell moved to Cambridge. One of his professors at Edinburgh commented in his recommendation letter, "He is not a little uncouth in manners, but withal one of the most original young men I have ever met with and with an extraordinary aptitude for physical enquiries."[23] Whereas the education at Edinburgh had been free-thinking and broad, Cambridge was far more competitive and intense, and much of his time was spent cramming for exams. After coming second in the university in his final exams, Maxwell was appointed as a Trinity College Fellow at the age of twenty-three. This gave him time to investigate a variety of phenomena, from fish-eye lenses to the flight of falling pieces of paper, and even the ability of cats to right themselves if dropped. He also demonstrated, using coloured spinning tops, that white light is a mixture of red, green, and blue.

Just a year later, in 1856, Maxwell moved to Aberdeen to take up a chair of natural philosophy. He spent five years there before moving to King's College, London. During this period he contributed to many different fields, applying in each case a deft combination of physical insight and mathematical skill. He showed that Saturn's rings were composed of particles, a theory confirmed by the *Voyager* flybys of the 1980s. He developed models of elasticity and discovered relations in the theory of heat, both of which are still used by engineers. Later on in his career, he worked out the statistical properties of molecules in a gas and he demonstrated the first-ever colour slide. But the feat that unquestionably trumps them all began in 1854, when he tried to clean up a bunch of messy equations having to do with electricity and magnetism.[24]

Michael Faraday, by contrast, was the son of a South London blacksmith and left school at thirteen to become a bookbinder's apprentice. He had no formal scientific education and no mathematics, but he had a deep curiosity about the world, an alertness to it, and marvellous physical intuition.

On reading an article on electricity in an encyclopedia he was binding, Faraday was captivated. One of the bookbinder's customers, perceiving the lad's evident intelligence and thirst for knowledge, gave him tickets to lectures by Sir Humphrey Davy, one of the great scientists of the day, at the Royal Institution. Having attended the lectures, Faraday copied out his copious notes, which amounted to a virtual transcription of the lectures, and presented them, beautifully bound, to the great man. This led to a job, first as a bottle washer in Davy's lab and, soon enough, as his right-hand man. Eventually he succeeded Davy as the director of the Royal Institution. Despite its walls of inequity and injustice, the Victorian age sometimes let in chinks of light, such as its workingmen's colleges and public lectures bringing science to the general populace.

As a mature scientist, Faraday was indefatigable and responsible for a staggering range of discoveries. But what fascinated him above all was electricity and magnetism, and he was by no means alone in this. Although electricity had been observed for millennia in certain shocking fish and in lightning, by the nineteenth century its magical properties were beginning to be widely appreciated, though they were not understood. Its spark and sizzle were lifelike — it galvanized the age, you might say. Mary Shelley's *Frankenstein; or, The Modern Prometheus* was inspired by electrical experiments, often carried out in public, on living and dead creatures in early nineteenth-century London. Its title compared the modern scientist to the ancient Greek hero Prometheus, a lesser god who became a champion of mankind. He stole fire from the king of the gods, Zeus, and gave it to man. Shelley's book was a cautionary tale: for his crime, Prometheus was condemned by Zeus to be chained to a rock and have his liver eaten out by an eagle every day, only for it to grow back every night.

Faraday came to know electricity better than anyone, and his work was far ahead of its time. He showed that chemical bonds are electrical, discovering the laws of electrolysis and electrical deposition of one metal onto another. Faraday had a genius for discovering new phenomena using simple experiments. He investigated the magnetic properties of bismuth, iodine, plaster of Paris, even blood and liver. He blew soap bubbles filled with various gases — oxygen, nitrogen, hydrogen — through a magnetized region. He found that an oxygen-filled bubble got stuck in the magnetized region because oxygen is paramagnetic. (The explanation had to wait another ninety years, for the invention of quantum mechanics.)

Faraday also demonstrated the process of electromagnetic induction: how you can seemingly pull electricity out of a magnet by moving a wire past it. Faraday employed this in his invention of dynamos and transformers, now used to generate and distribute electricity all over the world. He even discovered superionic conduction, the basic mechanism of modern fuel cells.[25]

Faraday also showed that when a metal container is electrically charged, the charge moves onto the outer surface. He sat in a square cage, twelve feet on a side, while his assistant charged it to 150,000 volts. Sparks flew wildly everywhere. His hair flared out in a halo, but he was unharmed — the charge was all on the outside. The next time you fly through a lightning storm in a plane, thank Michael Faraday for showing that it would be safe!

Faraday did much, much more besides, but for our story, the most vital contribution he made was to formulate, for the very first time, a strange, slippery concept that is central to modern physics: a field. Instead of electric charges attracting or repelling one another from a distance, Faraday believed there must be an intermediary that carried the influence of one charge to another.

Faraday was not mathematical, and he could develop the idea only through slow and difficult experiments. It would fall to theory to make the next breakthrough. The eccentric young Maxwell built Faraday's intuition into the most beautiful and powerful mathematical framework in physics, and in so doing solved one of the greatest enigmas of all time.

THE SIMPLEST FACT ABOUT electricity is that like charges repel and unlike charges attract. You can easily see this by taking a roll of brown plastic packing tape and sticking two long strips, sticky side down, side by side on a tabletop. Holding the end of one strip with your left hand and the other with your right hand, rip both strips off the table at once and allow them to hang vertically downwards. If you bring your hands together gently, the two strips will swing away from each other because of the electrical force of repulsion between them.

The opposite effect is seen with a small tweak of the experiment. Instead of laying the strips on the table separately, put one strip down first and lay the other directly on top of it. Now pull them off the table together, as a double strip. Neutralize the double strip by running your fingers gently down the tape (this has the effect of cancelling any net electric charge). And now, using both hands, rip the two strips apart from one end. Because there was no charge on the pair of strips to begin with, if one strip gains a charge, the other must have lost some. So if one is positive, the other must be negative. Bring the two strips together, one hanging down from each hand, and you will find they draw together.

Magnets, too, have ends that attract and repel: a positive end called the north pole and a negative end called the south pole. Two north poles or two south poles repel but a north and a south pole attract. The name comes about because Earth itself is a giant magnet, which causes the north pole of any magnet — like the needle of a compass — to point in a direction close to that of the Earth's north pole.

Faraday had come to picture electric and magnetic forces as operating through "lines of force." He had a real bee in his bonnet about these lines. He believed that forces could not be "felt at a distance," as Newton had described in the case of gravity. Newton's picture would mean that if you jiggled the sun up and down, then Earth, 93 million miles away from it, would have to jiggle in perfect synchrony. And other objects would have to jiggle, all the way to infinity.

For Faraday, the idea that forces were transmitted right across space instantaneously between two objects was ridiculous. Instead, he thought, there must be something that actually carries the forces through the space that separates them. Whatever that something is, it must surround a mass or a charge or a magnet at all times, even when there isn't an object for it to push. Faraday's brilliant insight was in fact the first glimpse of the concept of a "force field."

You are familiar with the north–south and east–west grid lines on a street map. Visualize, if you can, such a grid spreading through three-dimensional space, with its grid lines going north–south, east–west, and up–down, and with every grid line separated by the same distance from its neighbouring parallel lines. The grid is a convenient way of measuring length, width, and depth. Where any three grid lines cross, we have a grid point, labelled by its coordinate on each of the grid lines passing through it. We can make our grid finer and finer by making the spacing between the parallel lines as small as we like. For physicists, this picture of a grid is the way we convert our mental picture of space — as a three-dimensional entity — into numbers, so that each different point of space is associated with three numbers.[26]

To picture a force field, imagine attaching a little arrow to every grid point. Each arrow can point in any direction. The arrows represent the force field; the length and direction of an arrow indicate the strength and direction of the field at that point in space. Any charged particle placed in this force field will feel a force. For example, an electron feels a force given by its electric charge times the electric field.

When Maxwell was only twenty-five years old and working as a College Fellow in Cambridge, he wrote to Faraday, then one of the most famous scientists of the day and director of the Royal Institution. Maxwell enclosed a paper he had just written titled "On Faraday's Lines of Force," giving a mathematical description of the effects that Faraday had reported in his experiments. Faraday, who had never studied mathematics, later noted, "I was at first almost frightened, when I saw such mathematical force made to bear upon the subject."[27] Yet Maxwell wrote modestly in his paper, "By the method which I adopt, I hope to render it evident that I am not attempting to establish any physical theory of a science in which I have hardly made a single experiment, and that the limit of my design is to shew how, by a strict application of the ideas and methods of Faraday, the connection of very different orders of phenomena which he has discovered may be placed before the mathematical mind."[28] Their interaction and mutual respect illustrate beautifully the interplay between theory and experiment in its ideal form.

Faraday, encouraged by Maxwell's work, returned to his laboratory at the Royal Institution in London with renewed vigour. His goal now was to show that electric and magnetic fields take time to move through space. Already sixty-six years old and exhausted after many years of arduous experimentation, he did not succeed (it took three more decades before German physicist Heinrich Hertz finally did). It was time for Maxwell's "mathematical mind" to come to the rescue.

By the time Maxwell moved to King's College, London, in 1861, he was ready to begin making sense of electricity and magnetism in earnest. His goal was to describe mathematically the laws governing the "lines of force" envisaged by Faraday. In several stages, Maxwell built this intuition into a full-fledged theory of "fields," a concept that would dominate fundamental physics in the twentieth century.

Electric charges spew out electric fields, magnets spew out magnetic fields, and masses spew out gravitational fields so that all three kinds of fields are present everywhere in the universe. In modern terms, we represent electric and magnetic fields as a sea of little arrows filling space. The trick in describing all the equations of electricity and magnetism is to figure out how the arrows located at each point in space influence their neighbours. The rules are complicated, and Maxwell had to work them all out for the first time. Newton had invented calculus as the mathematics for describing motion. Maxwell had to extend this idea to describe how force fields change in space and in time. Using a grid of points in space, like the one we described, Maxwell developed the theory of partial differential equations to describe force fields. The mathematics he invented is used throughout science to describe fluids, the flow of air, or even the propagation of disease.

The way in which Maxwell found his equations for the electric and magnetic fields was at first sight surprising. He envisaged a machine whose moving parts represented the fields. In his first attempt, the lines of force of the electric field were represented by "tubes" carrying a fluid out of positive electric charges and into negative electric charges. But gradually the model became more sophisticated, and the fluid-filled "tubes" were replaced by microscopic "rollers" and "spinning wheels" representing both electric and magnetic fields, turning the whole of space into a gargantuan factory.

With guidance from William Thomson (Lord Kelvin), Maxwell laid out all the different phenomena and all the laws of electricity and magnetism known at the time — a veritable alphabet of laws: Ampère's, Biot-Savart's, Coulomb's, Faraday's, Franklin's, Gauss's, Kirchhoff's, Lenz's, Ohm's laws and more, built up over the course of the previous century. His goal was to fit all the pieces together into a single consistent mechanical framework.

Maxwell was able to incorporate each of the known laws of electricity and magnetism into his conceptual mechanism — except one. Benjamin Franklin had proposed that an electric charge could be neither created nor destroyed. Maxwell formulated Franklin's law mathematically, showing how the electric charge in a region of space is changed by the flow of electric current into or out of the region. Carl Friedrich Gauss had described how electric charges give rise to electric fields, and André-Marie Ampère had described how electric currents create magnetic fields. But when Maxwell put the three laws together, he found a contradiction: they were incompatible! The only way he could restore consistency was to change Ampère's law by adding a new term, according to which a changing electric field would also cause a magnetic field. This was, he realized, similar to the way in which a changing magnetic field generated an electric field, as Michael Faraday had demonstrated.

But wait: a changing electric field can now create a changing magnetic field which, by Faraday's law, can create a changing electric field. So *electric and magnetic fields can create one another, without any electric charges or currents or magnets being present.* When Maxwell analyzed his equations carefully, he found that magnetic and electric fields can travel across space together, like an undulating pattern moving across the grass in a meadow. This electromagnetic wave was just the kind of effect Faraday had been anticipating.

While summering in his ancestral home of Glenlair, in 1861, Maxwell made his discovery. Using the best experimental measurements to date, he worked out the speed at which the electromagnetic waves would travel. In a letter to Faraday he wrote, "The result is 193,088 miles per second (deduced from electrical and magnetic experiments). [French physicist Hippolyte] Fizeau has determined the velocity of light = 193,118 miles per second by direct experiment." And then, with lovely understatement, he added, "This coincidence is not merely numerical."[29]

Maxwell, using purely mathematical arguments, had not only predicted the speed of light, he had *explained light's nature.* Simply by piecing together known facts and insisting on mathematical consistency, he had revealed one of the most basic properties of the universe.

Once he had reached his conclusion, Maxwell swiftly set about scrapping his mechanical model. Now that he had the right equations, he no longer needed the visual machinery. The equations were the theory: one needed nothing more and nothing less.

Whenever I teach electromagnetism, Maxwell's discovery is the highlight of the course. There is a moment of sheer magic when the students suddenly see how all the pieces fit together and light has seemingly popped out of nowhere. "If you are ever in doubt," I tell them, "remember this moment. Perseverance leads to enlightenment!" And the truth is more beautiful than your wildest dreams.

IT WOULD BE HARD to overstate the importance of Maxwell's discovery in unifying electricity, magnetism, and light. All at once, this provided a simple, precise description of a vast array of phenomena: the spark from a brass knob on a cold morning; the signals that traverse our nerves or make our muscles move; lightning strikes and candlelight; the swing of a compass needle and the spin of an electric turbine.

The direct impact on technology would soon be felt in radio, television, and radar. But the long-term effect on basic physics was even greater. Maxwell's breakthrough opened the door to twentieth-century physics: to relativity, quantum theory, and particle physics, our most fundamental descriptions of reality.

One of the theory's most important predictions was that electromagnetic waves could have any wavelength, from zero to infinity. Just a tiny portion of this spectrum — from two-fifths to three-quarters of a micron (a millionth of a metre) — explains all of visible light: red, yellow, green, blue, violet. Maxwell's discovery widened the rainbow, predicting the existence of electromagnetic waves with wavelengths ranging from a thousandth the size of an atomic nucleus (the gamma rays produced in the Large Hadron Collider) to thousands of kilometres (the ultra-low-frequency waves used for communication with submarines). In between are X-rays, used for medical imaging; infrared waves, used for night vision; microwaves, used for cooking; and radio waves, used for everything from cellphones to telescopes probing neutron stars and black holes. Maxwell's equations describe every single one of these waves in exactly the same way. They are just stretched-out or shrunken-down versions of one another.

Maxwell's theory did much more. Slowly the realization dawned that it contradicted the two most hallowed frameworks in physics: Newton's theory of forces, motion, and gravity, and the equally firmly established theory of heat. As Maxwell's equations were studied, it was noticed by the Dutch physicist Hendrik Lorentz that they possessed a symmetry connecting space and time, which would open the door to Einstein's unification of space, time, mass, energy, and gravity, and plant the seed for the study of the entire evolving cosmos.

Likewise, Maxwell's theory opened the door to quantum theory. In trying to reconcile Maxwell's description of electromagnetic radiation with the theory of heat, Max Planck and then Albert Einstein discovered an inconsistency so drastic that, in time, it would overturn the entire classical picture of the world. Maxwell's theory emerged from this collision in a new form: as a quantum field theory. In taking this form, it set the pattern for all of twentieth-century physics.

We are still struggling with the implications of the quantum revolution. Our intuition is based on the classical picture of the world, a picture founded upon Newton's and Maxwell's discoveries, in which particles and fields have a definite existence and move around in space according to absolutely deterministic laws. But, as I will describe in the next chapter, that picture is gone, and a more mysterious, quantum conception of reality has emerged, incorporating a greater degree of possibility and giving us a greater role.

· · ·

I HAVE TITLED THIS work *From Quantum to Cosmos* because I want to celebrate with you our progress towards understanding nature at its most basic level. In subsequent chapters we will follow the journey physics has taken, from the quantum world to the cosmos to the unification of all known physics in a single equation. It is a story of fun, yearning, determination, and, most of all, humility and awe before nature.

Science is all about people. They may work in labs and scribble strange formulae, but they are driven by the same natural curiosity we are all born with: to explore and discover our world and what we can do. Some people are blessed with unusual mathematical abilities or physical insight; others make great discoveries through sheer persistence, careful planning, or just good luck. Science is, above all, a human activity. It is all about making the most of the marvellous gift of life.

When Usain Bolt smashed the world sprint records in Beijing in 2008, and again in Berlin in 2009, we all celebrated. Wasn't it fantastic to see seemingly impossible limits breached? In the same way, we should celebrate the even more remarkable achievements of Maxwell and Einstein and their modern counterparts. The world needs more people who are capable of making great discoveries, and they can come from anywhere. They are examples of our human nature and spirit, and we should all draw inspiration from their success.

Much of science is complicated and technical. Many of its ideas are difficult, but scientists can and must become much better at explaining what they are doing, and why. And society needs to appreciate far better how science brought us here, and where it might take us.

Reconnecting science to society has a deeper purpose than developing the next marketable technology. It is about the kind of society we want to create, a society in which there is optimism, confidence, and purpose. Scientists need to know why they are doing science, and society needs to know why it supports them.

The technologies we rely on today are all based on past discoveries. We need new breakthroughs and we need to find more intelligent ways of using the knowledge we already possess. The billions of young minds on our planet need to be carefully nurtured and encouraged. Each one is a potential Faraday, Maxwell, or Mandela, capable of transforming the world.

A new world is now beckoning. As I'll describe in the next chapter, quantum physics has revealed that the behaviour of the universe, and the way in which we are involved with it, is stranger than anyone could have expected. On the horizon are technologies and understanding beyond anything we have experienced so far. We are being challenged to rise to the next level of existence, the next stage in the evolution of ourselves and of the universe. Witnessing all the changes wrought by classical physics, we can only imagine what our quantum future holds — and what we will do with it.

TWO

OUR IMAGINARY REALITY

"Theoretical physicists live in a classical world,
looking out into a quantum-mechanical world."
— John Bell[1]

"Describing the physical laws without reference to
geometry is similar to describing our thoughts
without words."— Albert Einstein[2]

THE SCHOOL OF ATHENS, by Raphael, is one of the most
breathtaking paintings of the Italian Renaissance. It represents a key moment in human history: the flowering
of free thinking in Classical Greece. Somehow, the people of the time were able to look at the world with fresh
eyes, to set aside traditional superstitions and beliefs
in dogma or high authority. Rather, through discussion and logical argument, they began to figure out for
themselves how the universe works and what principles

human society should be based upon. In doing so, they changed Western history forever, forming many of the concepts of politics and literature and art that underlie the modern world.

Raphael's picture is full of philosophers like Aristotle, Plato, and Socrates engaged in discussion. There is also the philosopher Parmenides, in some ways the ancient Greek version of Stephen Hawking. Like Hawking, Parmenides believed that at its most fundamental level, the world is unchanging, whereas Heraclitus, also in the portrait, believed that the world is in ceaseless motion as a result of the tension between opposites. There are mathematicians too. At front right is Euclid, giving a demonstration of geometry, and at front left is Pythagoras, absorbed in writing equations in a big book. Beside Parmenides is Hypatia, the first woman mathematician and philosopher. The whole scene looks like a sort of marvellous university — which I, for one, would have loved to attend — full of people exploring, exchanging, and creating ideas.

An odd figure is peering over Pythagoras's shoulder and scribbling in a notebook. He looks as if he is cheating, and in some ways he is: he has one eye on the mathematics and the other on the real world. This is Anaximander, who some consider to be the world's first scientist.[3] He lived around 600 B.C. in Miletus, then the greatest Greek city, in the eastern Aegean on the coast of modern Turkey. At that time, the world was dominated by kings and traditional rulers with all kinds of mystical and religious beliefs. Yet somehow Anaximander, his teacher Thales, and his students — and the thinkers who followed — trusted their own powers of reason more than anyone had ever done before.

Almost every written record of their work has been lost, but what little we know is mind-boggling. Anaximander invented the idea of a map — hence quantifying our notion of space — and drew the first known map of the world. He is also credited with introducing to Greece the gnomon, an instrument for quantifying time, consisting of a rod set vertically in the ground so that its shadow showed the direction and altitude of the sun. Anaximander is credited with using the gnomon, in combination with his knowledge of geometry, to accurately predict the equinoxes.[4]

Anaximander also seems to have been the first to develop a concept of infinity, and he concluded, although we do not know how, that the universe was infinite. He also proposed an early version of biological evolution, holding that human beings and other animals arose from fish in the sea. Anaximander is considered the first scientist because he learned systematically from experiments and observations, as when he developed the gnomon. In the same way he was taught by Thales, he seems to have taught Pythagoras. Thus was built the scientific tradition of training students and passing along knowledge.

Just think of these phenomenal achievements for a moment, and imagine the transformations they eventually brought about. How often have you arrived in a strange city or neighbourhood without a map or a picture of your location? With nothing but your immediate surroundings, with no mental image of their context, you are lost. Each new turn you take brings something unexpected and unpredictable. A map brings a sense of perspective — you can anticipate and choose where you want to go and what you want to see. It raises entirely new questions. What lies beyond the region shown in the map? Can we map the world? And the universe?

And how would you think of time without a clock? You could still use light and dark, but all precision would be lost in the vagaries of the seasons and the weather. You would live far more in the present, with the past and the future being blurred. The measurement of time opened the way to precise technologies, like tracking and predicting celestial bodies and using them as a navigational tool. Yet even these matters were probably not Anaximander's primary concerns. He seems to have been more interested in big questions, such as what happened if you traced time back into the distant past or far forward into the future.

What about Anaximander's idea that the universe is infinite? This seems plausible to us now, but I distinctly remember that when I was four years old, I thought the sky was a spherical ceiling, with the sun and stars fixed upon it. What a change it was when I suddenly realized, or was told, that we are looking out into an infinite expanse of space. How did Anaximander figure that out? And what about his idea that we arose from fish in the sea? His ideas suggested above all a world with potential. If things were not always as they are now, they might be very different in the future.

It was no accident that these beginnings of modern science occurred around the same time as many new technologies were being invented. Nearby, on the island of Samos, Greek sculptor and architect Theodorus developed, or at least perfected, many of the tools of the building trade: the carpenter's square, the water level, the turning lathe, the lock and key, and the craft of smelting.

Hand in hand with these developments was a flowering of mathematics, philosophy, art, literature, and, of course, democracy. But the civilization of ancient Greece was fleeting. Throughout its existence, it was ravaged by wars and invasions: the Greco-Persian Wars, the war between Athens and Sparta, the invasion by Alexander the Great, and the chaos following his death. Finally, there was the triumph of Rome and then its decadent decline, which snuffed out civilization in Europe for a millennium. The great libraries of the ancient world, like the one at Alexandria, were lost. Only fragments and copies of their collections survived.

In the fifteenth century, Aldus Manutius, Italy's leading printer, made it his personal mission to reproduce cheap and accurate pocket editions of the ancient classics and make them widely available. In this way, the ideas of ancient Greece directly seeded the Renaissance and the Scientific Revolution that followed.

MORE THAN FOUR HUNDRED years later, we come to a modern counterpart of Raphael's masterpiece: a black-and-white photograph of the Fifth Solvay International Conference on Electrons and Photons, held in Brussels in 1927.

Towards the end of the nineteenth century, physicists had felt they were close to converging on a fundamental description of nature. They had Newton's laws of mechanics; Maxwell's theory of electricity, magnetism, and light; and a very successful theory of heat founded by Maxwell's friend William Thomson (Lord Kelvin), among others. Physics had provided the technical underpinning of the Industrial Revolution and had opened the way to global communication. A few small details remained to be wrapped up, like the inner structure of the atom. But the classical picture of a world consisting of particles and fields moving through space and time seemed secure.

Instead, the early twentieth century brought a series of surprises. The picture became increasingly confused and was only resolved by a full-scale revolution between 1925 and 1927. In this revolution, the physicists' view of the universe as a kind of large machine was completely overturned and replaced by something far less intuitive and familiar. The Fifth Solvay Conference was convened just as this new and abstract representation of the world had formed. It might be considered the most uncomfortable conference ever held in physics.

In 1925, the young German prodigy Werner Heisenberg launched quantum theory with a call to "discard all hope of observing hitherto unobservable quantities, such as the position and period of the electron," and instead to "try to establish a theoretical quantum mechanics, analogous to classical mechanics, but in which only relations between observable quantities occur."[5] Heisenberg's work replaced the classical picture of the electron orbiting the atomic nucleus with a far more abstract, mathematical description, in which only those quantities that were directly observable in experiments would have any literal interpretation. Soon after, in 1926, the Austrian physicist Erwin Schrödinger found an equivalent description to Heisenberg's, in which the electron was treated as a wave instead of a classical particle. Then, in early 1927, Heisenberg discovered his famous uncertainty principle, showing that the central concept in Newton's classical universe — that every particle had a definite position and velocity — could not be maintained.

By the time the physicists got to Brussels for the Solvay Conference, the classical view of the world had finally collapsed. They had to give up any notion of making definite predictions because there was, in a sense, no longer a definite world at all. As Max Born had realized in 1926, quantum physics could only make statements about probabilities. But it wasn't even a case of little demons playing dice in the centre of atoms: it was far stranger than that. There was an elegant mathematical formalism governing the world's behaviour, but it had no classical interpretation. No wonder all the physicists at Solvay are looking so glum!

Front and centre in the photo, of course, is Einstein. Next to him, with his legs crossed, is Dutch physicist Hendrik Lorentz. And then there is Marie Curie — the only woman in the picture and also the only one among them to win *two* Nobel prizes. Curie, with her husband Pierre, had shown that radioactivity was an atomic phenomenon. Their discovery was one of the first hints of the strange behaviour in the subatomic world: radioactivity was finally explained, a year after the Fifth Solvay meeting, as the quantum mechanical tunnelling of particles out of atomic nuclei. Next to Curie is Max Planck, holding his hat and looking sad. Planck had been responsible for initiating the quantum revolution in 1900 with his suggestion that light carries energy in packets called "photons." His ideas had been spectacularly confirmed in 1905, when Einstein developed them to explain how light ejects electrons from metals.

In the middle of the next row back, between Lorentz and Einstein, is Paul Dirac, the English genius and founder of modern particle physics, with Erwin Schrödinger standing behind him. Werner Heisenberg is standing at the back, three in from the right, with the German-British mathematician Max Born sitting in front of him. Heisenberg and Born had together developed the matrix mechanics formulation of quantum theory, which Dirac had brought to its final mathematical form. Next to Born is the Danish physicist Niels Bohr, a towering figure who had extended Planck's quantum idea to the hydrogen atom and who had since played

the role of godfather to quantum theory. Bohr founded
the Institute for Theoretical Physics of the University of
Copenhagen and became its director. There, he mentored
Heisenberg and many other physicists; it became a world
centre for quantum theory. Heisenberg would later say,
"To get into the spirit of the quantum theory was, I would
say, only possible in Copenhagen at that time [1924]."[6]
Bohr was responsible for developing what became the
most popular interpretation of quantum theory, known
as the Copenhagen interpretation. On Heisenberg's right
is Wolfgang Pauli, the young Austrian prodigy who had
invented the Pauli exclusion principle, stating that two
electrons could not be in the same state at the same
time. This principle, along with the quantum theory of
spin that Pauli also developed, proved critical in understanding how electrons behave within more complicated
atoms and molecules. Dirac, Heisenberg, and Pauli were
only in their mid-twenties and yet at the forefront of the
new developments.

The participants came from a wide range of backgrounds. Curie was more or less a refugee from Poland.[7]
Einstein himself had worked at a patent office before
making his sensational discoveries in 1905. He, Born,
Bohr, their great friend Paul Ehrenfest (standing behind
Curie in the photo), and Pauli's father were represen-
tatives of a generation of young Jews who had entered
maths and science in the late nineteenth century. Before
that time, Jews had been deliberately excluded from
universities in western Europe. When they were finally

allowed to enter physics, maths, and other technical fields, they did so with a point to prove. They brought new energy and ideas, and they would dispel forever any notion of intellectual inferiority.

So there we have many of the world's leading physicists meeting to contemplate a revolutionary new theory — and to figure out its repercussions for our view of the universe. But they seemed none too happy about it. They had discovered that, at a fundamental level, the behaviour of nature's basic constituents is truly surreal. They just don't behave like particles or billiard balls or masses sliding down planes, or weights on springs or clouds or rivers or waves or anything anyone has ever seen in everyday life. Even Heisenberg saw the negative side: "Almost every progress in science has been paid for by a sacrifice; for almost every new intellectual achievement, previous positions and conceptions had to be given up. Thus, in a way, the increase of our knowledge and insight diminishes continually the scientist's claim on 'understanding' nature."[8] On the other hand, objectively speaking, the 1920s were a golden age for physics. Quantum theory opened up vast new territories where, as Dirac told me when I met him many years later, "even mediocre physicists could make important discoveries."

It would take most of the remainder of the twentieth century for physicists to fully appreciate the immense opportunities that quantum physics offers. Today, we stand on the threshold of developments through which it may completely alter our future.

. . .

THE STRANGE STORY OF the quantum begins with the humble electric light bulb. In the early 1890s, Max Planck, then a professor in Berlin, was advising the German Bureau of Standards on how to make light bulbs more efficient so that they would give out the maximum light for the least electrical power. Max Born later wrote about Planck: "He was by nature and by the tradition of his family conservative, averse to revolutionary novelties and skeptical towards speculations. But his belief in the imperative power of logical thinking based on facts was so strong that he did not hesitate to express a claim contradicting to all tradition, because he had convinced himself that no other resort was possible."[9]

Planck's task was to predict how much light a hot filament gives out. He knew from Maxwell's theory that light consists of electromagnetic waves, with each wavelength describing a different colour of light. He had to figure out how much light of each colour a hot object emits. Between 1895 and 1900, Planck made a series of unsuccessful attempts. Eventually, in what he later called an "act of despair,"[10] he more or less worked backward from the data, inferring a new rule of physics: that light waves could accept energy only in packets, or "quanta." The energy of a packet was given by a new constant of nature, Planck's constant, times the oscillation frequency of the light wave: the number of times per second the electric and magnetic fields vibrate back and forth as an electromagnetic wave travels past any point in space. The oscillation frequency is given by the speed of light divided by the wavelength of the light. Planck found that with this rule he could perfectly match the experimental measurements of the spectrum of light emitted from hot objects. Much later, Planck's energy packets became known as photons.

Planck's rule was less *ad hoc* than it might at first seem. He was a sophisticated theorist, and well appreciated a powerful formalism that had been developed by the Irish mathematical physicist William Rowan Hamilton in the 1830s, building on earlier ideas of Fermat, Leibniz, and Maupertuis. Whereas Newton had formulated his laws of motion as rules for following a system forward from one moment in time to the next, Hamilton considered all the possible histories of a system, from some initial time to some final time. He was able to show that the actual history of the system, the one that obeyed Newton's laws, was the one that minimized a certain quantity called the "action."

Let me try to illustrate Hamilton's idea with the example of what happens when you're leaving a supermarket. When you're done with your grocery shopping, you're faced with a row of checkouts. The nearest will take you less time to walk to, but there may be more people lined up. The farthest checkout will take longer to walk to but may be empty. You can look to see how many people have baskets or trolleys, how much stuff is in them, and how much is on the belt. And then you choose what you think will be the fastest route.

This is, roughly speaking, the way Hamilton's principle works. Just as you minimize the time it takes to leave the supermarket, physical systems evolve in time in such a way as to minimize the action. Whereas Newton's laws describe how a system edges forward in time, Hamilton's method surveys *all* the available paths into the future and chooses the best among them.

Hamilton's new formulation allowed him to solve many problems that could not be solved before. But it was much more than a technical tool: it provided a more integrated picture of reality. It helped James Clerk Maxwell develop his theory of electromagnetism, and it guided Planck to an inspired guess that launched quantum theory. In fact, Hamilton's version of mechanics had anticipated the future development of quantum theory. Just as you find when leaving the supermarket that there may be several equally good options, Hamilton's action principle suggests that in some circumstances the world might follow more than one history. Planck was not ready to contemplate such a radical departure from physics' prior perspectives but, decades later, others would. As we will see in Chapter Four, by the end of the twentieth century all the known laws of physics were expressed in terms of the quantum version of Hamilton's action principle.

The point of all this for our story is that Planck knew that Hamilton's action principle was a fundamental formulation of physics. It was therefore natural for him to try to relate his quantization rule to Hamilton's action.

The units in which Hamilton's action is measured are energy times time. The only time involved in a light wave is the oscillation period, equal to the inverse of the oscillation frequency. So Planck guessed that the energy of an electromagnetic wave times its oscillation period is equal to a whole-number multiple of a new constant of nature, which he called the "action quantum" and which we now call "Planck's constant." Because Planck believed that all the laws of physics could be included in the action, he hoped that one day his hypothesis of quantization might become a new universal law. In this guess, he would eventually be proven right.

Two of Planck's colleagues at Berlin, Ferdinand Kurlbaum and Heinrich Rubens, were leading experimentalists of the time. By 1900, their measurements of the spectrum of light emitted from hot objects had become very accurate. Planck's new guess for the spectrum, based on his quantization rule, fitted their data beautifully and explained the changes in colour as an object heats up. For this work, Planck came to be regarded as the founder of quantum theory. He tried but failed to go further. He later said: "My unavailing attempts to somehow reintegrate the action quantum into classical theory extended over several years and caused me much trouble."[11] Physics had to wait for someone young, bold, and brilliant enough to make the next leap.

PLANCK WAS GUIDED TO his result in part by theory and in part by experiment. In 1905, Albert Einstein published a clearer and more incisive theoretical analysis of why the classical description of electromagnetic radiation failed to describe the radiation from hot objects.

The most basic notion in the theory of heat is that of thermal equilibrium. It describes how energy is shared among all the components of a physical system when the system is allowed to settle down. Think of an object that, when cool, is perfectly black in colour, so it absorbs any light that falls on it. Now heat up this object and place it inside a closed, perfectly insulating cavity. The hot object emits radiation, which bounces around inside the cavity until it is reabsorbed. Eventually, an equilibrium will be established in which the rate at which the object emits energy — the quantity Planck wanted to predict — equals the rate at which it absorbs energy. In equilibrium, there must be a perfect balance between emission and absorption, at every wavelength of light. So it turns out that in order to work out the rate of emission of light from a perfectly black object when it is hot, all you need to know is how much radiation of each wavelength there is inside a hot cavity, which has reached equilibrium.

The Austrian physicist Ludwig Boltzmann had developed a statistical method for describing thermal equilibrium. He had shown that in many physical systems, on average, the energy would be shared equally among every component. He called this the "principle of equipartition." Einstein realized that electromagnetic waves in a cavity should follow this rule, and that this created a problem for the classical theory. The trouble was that

Maxwell's theory allows electromagnetic waves of all wavelengths, down to zero wavelength. There are only so many ways to fit a wave of a given wavelength inside a cavity of a certain size. But for shorter and shorter waves, there are more and more ways to fit the waves in. When we include waves of *arbitrarily* short wavelength, there are *infinitely* many different ways to arrange them inside the cavity. According to Boltzmann's principle, every one of these arrangements will carry the same average amount of energy. Together, they have an *infinite capacity to absorb heat*, and they will, if you let them, soak up all the energy.

Again, let me try to draw an analogy. Think of a country whose economy has a fixed amount of money in circulation (not realistic, I know!). Imagine there are a fixed number of people, all buying and selling things to and from each other. If the people are all identical (not realistic, either!), we would expect a law of *equipartition* of money. On average, everyone would have the same amount of money: the total amount of money divided by the total number of people.

Now imagine introducing more, smaller people into the country. For example, introduce twice as many people of half the size, four times as many a quarter the size, eight times as many people one-eighth the size, and so on. You just keep adding people, down to zero size, with all of them allowed to buy and sell in exactly the same way. Now, I hope you can see the problem: if you let the tiny people trade freely, because there are so many of them they will absorb all the money and leave nothing for anyone else.

Planck's rule is like imposing an extra "market reg-ulation" stating that people can trade money only in a certain minimum quantum, which depends inversely on their size. Larger people can trade in cents. People half as big can trade only in amounts of two cents, people half as big again in four cents, and so on. Very small people can trade only in very large amounts — they can buy and sell only very expensive, luxury items. And the smallest people cannot trade at all, because their money quantum would be larger than the entire amount of money in circulation.

With this market regulation rule, an equilibrium would be established. Smaller people are more numerous and have a larger money quantum. So there is a certain size of people that can share all the money between them, and still each have enough for a few quanta so they aren't affected by the market regulation. In equilibrium, people of this size will hold most of the money. Increase the total money in circulation, and you will decrease the size of the people holding most of the money.

In the same way, Einstein showed, if you imposed Planck's quantization rule, most of the energy inside a hot cavity would be held by waves just short enough to each hold a few quanta while sharing all the energy between them. Heat up the cavity, and shorter and shorter waves will share the energy in this way. Therefore if a hot body is placed inside the cavity and allowed to reach equilibrium, the wavelength of radiation it emits and absorbs decreases as the cavity heats up.

And this is exactly how hot bodies behave. If you heat up an object like a metal poker, as it gets hotter it glows red, then yellow, then white, and finally blue and violet when it becomes extremely hot. These changes are due to the decrease in wavelength of the light emitted, predicted by Planck's quantization rule. Human beings have been heating objects in fires for millennia. The colour of the light shining at them was telling them about quantum physics all along.

In fact, as we understand physics today, it is only Planck's quantization rule that prevents the short wavelength electromagnetic waves from dominating the emission of energy from any hot object, be it a lighted match or the sun. Without Planck's "market regulation," the tiniest wavelengths of light would be like the "Dementors" in the *Harry Potter* books, sucking all the energy out of everything else. The disaster that Planck avoided is referred to as the "ultraviolet catastrophe" of classical physics, because the shortest wavelengths of visible light are violet. (In this context, "ultraviolet" just means "very short wavelength.")

It is tempting to draw a parallel between this ultraviolet catastrophe in classical physics and what is now happening in our modern digital age. As computers and the internet become increasingly powerful and cheap, the ability to generate, copy, and distribute writing, pictures, movies, and music at almost no cost is creating another ultraviolet catastrophe, an explosion of low-grade, useless information that is in danger of overwhelming any valuable content. Where will it all end? Digital processors are now becoming so small that over the next decade they will approach the limits imposed by the size of atoms. Operating on these scales, they will no longer behave classically and they will have to be accessed and operated in quantum ways. Our whole notion of information will have to change, and our means of creating and sharing it will become much closer to nature. And in nature, the ultraviolet catastrophe is avoided through quantum physics. As I will discuss in the last chapter, quantum computers may open entirely new avenues for us to experience and understand the universe.

EINSTEIN'S 1905 PAPER CLEARLY described the ultraviolet catastrophe in classical physics and how Planck's quantum rule resolved it. But it went much farther, showing that the quantum nature of light could be independently seen through a phenomenon known as the "photoelectric effect." When ultraviolet light is shone on the surface of a metal, electrons are found to be emitted. In 1902, the German physicist Philipp Lenard had studied this phenomenon and showed that the energy of the individual emitted electrons increased with the frequency of the light. Einstein showed that the data could be explained if the electrons were absorbing light in quanta, whose energy was given by Planck's rule. In this way, Einstein found direct confirmation of the quantum hypothesis.

Yet, like Planck, Einstein also saw the worrying implications of quantization for any classical view of reality. He was later quoted as saying: "It was as if the ground was pulled out from under one."[12]

In 1913, the upheaval continued when Niels Bohr, working at Manchester under Ernest Rutherford, published a paper titled "On the Constitution of Atoms and Molecules." Much as Planck had done for light, Bohr invoked quantization to explain the orbits of electrons in atoms. Just before Bohr's work, Rutherford's experiments had revealed the atom's inner structure, showing it to be like a miniature solar system, with a tiny, dense nucleus at its centre and electrons whizzing around it.

Rutherford used the mysterious alpha particles, which Marie and Pierre Curie had observed to be emitted from radioactive material, as a tool to probe the structure of the atom. He employed a radioactive source to bombard a thin sheet of gold foil with alpha particles, and he detected how they scattered. He was amazed to find that most particles went straight through the metal but a few bounced back. He concluded that the inside of an atom is mostly empty space, with a tiny object — the atomic nucleus — at its centre. Legend has it that the morning after Rutherford made the discovery, he was scared to get out of bed for fear he would fall through the floor.[13]

Rutherford's model of the atom consisted of a tiny, positively charged nucleus orbited by negatively charged electrons. Since unlike charges attract, the electrons are drawn into orbit around the nucleus. However, according to Maxwell's theory of electromagnetism, as the charged electrons travelled around the nucleus they would cause changing electric and magnetic fields and they would continually emit electromagnetic waves. This loss of energy would cause the electrons to slow down and spiral inward to the nucleus, causing the atom to collapse. This would be a disaster every bit as profound as the ultraviolet catastrophe: it would mean that every atom in the universe would collapse in a very short time. The whole situation was very puzzling.

Niels Bohr, working with Rutherford, was well aware of the puzzle. Just as Planck had quantized electromagnetic waves, Bohr tried to quantize the orbits of the electron in Rutherford's model. Again, he required that a quantity with the same units as Hamilton's action — in Bohr's case, the momentum of the electron times the circumference of its orbit — came in whole-number multiples of Planck's constant. A hydrogen atom is the simplest atom, consisting of just one electron in orbit around a proton, the simplest nuclear particle. One quantum gave the innermost orbit, two the next orbit, and so on. As Bohr increased the number of quanta, he found his hypothesis predicted successive orbits, each one farther from the nucleus. In each orbit, the electron has a certain amount of energy. It could "jump" from one orbit to another by absorbing or emitting electromagnetic waves with just the right amount of energy.

Experiments had shown that atoms emitted and absorbed light only at certain fixed wavelengths, corresponding through Planck's rule to fixed packets of energy. Bohr found that with his simple quantization rule, he could accurately match the wavelengths of the light emitted and absorbed by the hydrogen atom.

· · ·

PLANCK, EINSTEIN, AND BOHR'S breakthroughs had revealed the quantum nature of light and the structure of atoms. But the quantization rules they imposed were *ad hoc* and lacked any principled basis. In 1925, all that changed when Heisenberg launched a radically new view of physics with quantization built in from the start. His approach was utterly ingenious. He stepped back from the classical picture, which had so totally failed to make sense of the atom. Instead, he argued, we must build the theory around the only directly observable quantities — the energies of the light waves emitted or absorbed by the orbiting electrons. So he represented the position and momentum of the electron in terms of these emitted and absorbed energies, using a technique known as "Fourier analysis in time."

At the heart of Fourier's method is a strange number called i, the imaginary number, the square root of minus one. By definition, i times i is minus one. Calling i "imaginary" makes it sound made up. But within mathematics i is every bit as definite as any other number, and the introduction of i, as I shall explain, makes the numbers more complete than they would otherwise be. Before

Heisenberg, physicists thought of *i* as merely a convenient mathematical trick. But in Heisenberg's work, *i* was far more central. This was the first indication of reality's imaginary aspect.

The imaginary number *i* entered mathematics in the sixteenth century, during the Italian Renaissance. The mathematicians of the time were obsessed with solving algebraic equations. Drawing on the results of Indian, Persian, and Chinese mathematicians before them, they started to find very powerful formulae. In 1545, Gerolamo Cardano summarized the state of the art in algebra, in his book *Ars Magna* (*The Great Art*). He was the first mathematician to make systematic use of negative numbers. Before then, people believed that only positive numbers made sense, since one cannot imagine a negative number of objects or a negative distance or negative time. But as we all now learn in school, it is often useful to think of numbers as lying on a number line, running from minus infinity to plus infinity from left to right, with zero in the middle. Negative numbers can be added, subtracted, multiplied, or divided just as well as positive numbers can.

Cardano and others had found general solutions to algebraic equations, but sometimes these solutions involved the square root of a negative number. At first sight, they discarded such solutions as meaningless. Then Scipione del Ferro invented a secret method of pretending these square roots made sense. He found that by manipulating the formulae he could sometimes get these square roots to cancel out of the final answer, allowing him to find many more solutions of equations.

There was a great deal of intrigue over this trick, because the mathematicians of the time held public contests, sponsored by wealthy patrons, in which any advantage could prove lucrative. But eventually the trick was published, first by Cardano and then more completely by Rafael Bombelli. In his 1572 book, simply titled *Algebra*, Bombelli systematically explained how to extend the rules of arithmetic to include i.[14] You can add, subtract, multiply, or divide it with any ordinary number. When you do, you will obtain numbers like $x + iy$, where x and y are ordinary numbers. Numbers like this, which involve i, are called "complex numbers." Just as we can think of the ordinary numbers as lying on a number line running from negative to positive values, we can think of the complex numbers as lying in a plane, where x and y are the horizontal and vertical coordinates. Mathematicians call this the "complex plane." The number zero is at the origin and any complex number has a squared length, given by Pythagoras's rule as $x^2 + y^2$.

Then it turns out, rather beautifully, that any complex number raised to the power of any other complex number is also a complex number. There are no more problems with square roots or cube roots or any other roots. In this sense, the complex numbers are *complete*: once you have added i, and any multiple of i, to the ordinary numbers, you do not need to add anything else. And later on, mathematicians proved that when you use

complex numbers, *every* algebraic equation has a solution. This result is called the "fundamental theorem of
algebra." To put it simply, the inclusion of *i* makes algebra
a far more beautiful subject than it would oterhwise be.

And from this idea came an equation that Richard
Feynman called "the most remarkable formula in mathematics."[15] It was discovered by Leonhard Euler, one of
the most prolific mathematicians of all time. Euler was
the main originator and organizer of the field of analysis — the collection of mathematical techniques for
dealing with infinities. One of his many innovations
was his use of the number *e*, which takes the numerical value 2.71828...and which arises in many areas of
mathematics. It describes exponential growth and is
used in finance for calculating compound interest or the
cumulative effects of economic inflation, in biology for
the multiplication of natural populations, in information science, and in every area of physics. What Euler
found is that *e* raised to *i* times an angle gives the two
basic trigonometric functions, the sine and cosine. His
formula connects algebra and analysis to geometry. It is
used in electrical engineering for the flow of AC currents
and in mechanical engineering to study vibrations; it is
also used in music, computer science, and even in cosmology. In Chapter Four, we shall find Euler's formula at
the heart of our unified description of all known physics.

Heisenberg used Euler's formula (in the form of a
Fourier series in time) to represent the position of an
electron as a sum of terms involving the energy states

of the atom. The electron's position became an infinite array of complex numbers, with every number representing a connection coefficient between two different energy states of the atom.

The appearance of Heisenberg's paper had a dramatic effect on the physicists of the time. Suddenly there was a mathematical formalism that explained Bohr's rule for quantization. However, within this new picture of physics, the position or velocity of the electron was a complex matrix, without any familiar or intuitive interpretation. The classical world was fading away.

Not long after Heisenberg's discovery, Schrödinger published his famous wave equation. Instead of trying to describe the electron as a point-like particle, Schrödinger described it as a wave smoothly spread out over space. He was familiar with the way in which a plucked guitar string or the head of a drum vibrates in certain specific wave-like patterns. Developing this analogy, Schrödinger found a wave equation whose solutions gave the quantized energies of the orbiting electron in the hydrogen atom, just as Heisenberg's matrices had done. Heisenberg's and Schrödinger's pictures turned out to be mathematically equivalent, though most physicists found Schrödinger's waves more intuitive. But, like Heisenberg's matrices, Schrödinger's wave was a complex number. What on earth could it represent?

Shortly before the Fifth Solvay Conference, Max Born proposed the answer: Schrödinger's wavefunction was a "probability wave." The probability to find the particle at any point in space is the squared length of the wavefunction in the complex plane, given by the Pythagorean theorem. In this way, geometry appeared at the heart of quantum theory, and the weird complex numbers that Heisenberg and then Schrödinger had introduced became merely mathematical tools for obtaining probabilities.

This new view of physics was profoundly dissatisfying to physicists like Einstein, who wanted to visualize concretely how the world works. In the run-up to the Solvay meeting, all hope of that was dashed. Heisenberg published his famous uncertainty principle, showing that, within quantum theory, you could not specify the position and velocity of a particle at the same time. As he put it, "The more precisely the position [of an electron] is determined, the less precisely the momentum is known in this instant, and vice versa."[16] If you know *exactly* where a particle is now, then you cannot say *anything* about where it will be a moment later. The very best you can hope for is a fuzzy view of the world, one where you know the position and velocity approximately.

Heisenberg's arguments were based on general principles, and they applied to any object, even large ones like a ball or a planet. For these large objects, the quantum uncertainty represents only a tiny ambiguity in their position or velocity. However, as a matter of principle, the uncertainty is always there. What Heisenberg's uncertainty principle showed is that, in quantum theory, nothing is as definite as Newton, or Maxwell, or any of the pre-quantum physicists had supposed it to be. Reality is far more slippery than our classical grasp of it would suggest.

ONE OF THE MOST beautiful illustrations of the quantum nature of reality is the famous "double-slit experiment." Imagine placing a partition with two narrow, parallel slits in it, between a source of light of one colour—like a green laser—and a screen. Only the light that falls on a slit will pass through the partition and travel on to the screen. The light from each slit spreads out through a process called "diffraction," so that each slit casts a broad beam of light onto the screen. The two beams of light overlap on the screen.

However, the distance the light has to travel from either slit to each point on the screen is in general different, so that when the light waves from both slits arrive at the screen, they may add or they may cancel. The pattern of light formed on the screen is called an "interference pattern": it consists of alternate bright and dark stripes at the locations where the light waves from the two slits add or cancel.[17] Diffraction and interference are classic examples of wave-like behaviour, seen not only in light but in water waves, sound waves, radio waves, and so on.

Now comes the quantum part. If you dim the light source and replace the screen with a detector, like a digital camera sensitive enough to detect individual photons—Planck's quanta of light—then you can watch the individual photons arrive. The light does not arrive as a continuous beam with a fixed intensity. Instead, the photons arrive as a random string of energy packets, each one announcing its arrival at the camera with a flash. The pattern of flashes still forms interference stripes, indicating that even though each photon of light arrived in only one place as an energy packet, the photons travelled through both slits and interfered as waves.

Now comes the strangest part. You can make the light source so dim that the long interval between the flashes on the screen means there is never more than one photon in the apparatus at any one time. But then, set the camera to record each separate flash and add them all up into a picture. What you find is that the picture *still* consists of interference stripes. Each individual photon interfered with itself, and therefore must somehow have travelled through *both* slits on the way to the screen.

So we conclude that photons sometimes behave like particles and sometimes behave like waves. When you detect them, they are always in a definite position, like a particle. When they travel, they do so as waves, exploring all the available routes; they spread out through space, diffract, and interfere, and so on.

In time, it was realized that quantum theory predicts that electrons, atoms, and every other kind of particle also behave in this way. When we detect an electron, it is always in a definite position, and we find all its electric charge there. But when it is in orbit around an atom, or travelling freely through space, it behaves like a wave, exhibiting the same properties of diffraction and interference.

In this way, quantum theory unified forces and particles by showing that each possessed aspects of the other. It replaced the world that Newton and Maxwell had developed, in which particles interacted through forces due to fields, with a world in which both the particles and the forces were represented by one kind of entity: quantized fields possessing both wave-like and particle-like characters.

NIELS BOHR DESCRIBED THE coexistence of the wave and particle descriptions with what he called the "principle of complementarity." He posited that some situations were best described by one classical picture — like a particle — while other situations were better described by another — like a wave. The key point was that there was no logical contradiction between the two. The words of the celebrated American author of the time, F. Scott Fitzgerald, come to mind: "The test of a first-rate intelligence is the ability to hold two opposed ideas in the mind at the same time, and still retain the ability to function."[18]

Bohr had a background in philosophy as well as mathematics, and an exceptionally agile and open mind. His writings are a bit mystical and also somewhat impenetrable. His main role at the Solvay Conference seems to have been to calm everyone down and reassure them that despite all the craziness everything was going to work out fine. Somehow, Bohr had a very deep insight that quantum theory was consistent. It's clear he couldn't prove it. Nor could he convince Einstein.

Einstein was very quiet at the Fifth Solvay meeting, and there are few comments from him in the recorded proceedings. He was deeply bothered by the random, probabilistic nature of quantum theory, as well as the abstract nature of the mathematical formalism. He famously remarked (on a number of occasions), "God does not play dice!" To which at some point Bohr apparently replied, "Einstein, stop telling God how to run the world."[19] At this and subsequent Solvay meetings, Einstein tried again and again to come up with a paradox that would expose quantum theory as inconsistent or incomplete. Each time, after a day or two's thought, Bohr was able to resolve the paradox.

Einstein continued to be troubled by quantum theory, and in particular by the idea that a particle could be in one place when it was measured and yet spread out everywhere when it was not. In 1935, with Boris Podolsky and Nathan Rosen, he wrote a paper that was largely ignored by physicists at the time because it was considered too philosophical. Nearly three decades later, it would seed the next revolutionary insight into the nature of quantum reality.

Einstein, Podolsky, and Rosen's argument was ingenious. They considered a situation in which an unstable particle, like a radioactive nucleus, emits two smaller, identical particles, which fly apart at exactly the same speed but in opposite directions. At any time they should both be equidistant from the point where they were both emitted. Imagine you let the two particles get very far apart before you make any measurement — for the sake of argument, make it light years. Then, at the very last minute, as it were, you decide to measure either the position or the velocity of one of the particles. If you measure its position, you can infer the position of the other without measuring it at all. If instead you measure the velocity, you will know the velocity of the other particle, again without measuring it. The point was that you could decide whether to measure the position or the velocity of one particle when the other particle was so far away that it could not possibly be influenced by your decision. Then, when you made your measurement, you could infer the second particle's position or velocity. So, Einstein and his colleagues argued, the unmeasured particle must really have both a position and a velocity, even if quantum theory was unable to describe them both at the same time. Therefore, they concluded, quantum theory must be incomplete.

Other physicists balked at this argument. Wolfgang Pauli said, "One should no more rack one's brain about the problem of whether something one cannot know anything about exists all the same, than one should about the ancient question of how many angels are able to sit on the point of a needle."[20] But the Einstein–Podolsky–Rosen argument would not go away, and in the end someone saw how to make use of it.

· · · · ·

HAVE YOU EVER WONDERED whether there is a giant conspiracy in the world and whether things really are as they appear? I'm thinking of something like the *The Truman Show*, starring Jim Carrey as Truman, whose life appears normal and happy but is in fact a grand charade conducted for the benefit of millions of TV viewers. Eventually, Truman sees through the sham and escapes to the outside world through an exit door in the painted sky at the edge of his arcological dome.

In a sense, we all live in a giant Truman show: we conceptualize the world as if everything within it has definite properties at each point in space and at every moment of time. In 1964, the Irish physicist John Bell discovered a way to show conclusively that any such classical picture could, with some caveats, be experimentally disproved.

Quantum theory had forced physicists to abandon the idea of a deterministic universe and to accept that the best they could do, even in principle, was to predict probabilities. It remained conceivable that nature could be pictured as a machine containing some hidden mechanisms that, as Einstein put it, threw dice from time to time. One example of such a theory was invented by the physicist David Bohm. He viewed Schrödinger's wavefunction as a "pilot wave" that guided particles forward in space and time. But the actual locations of particles in his theory are determined statistically, through a physical mechanism to which we have no direct access. Theories that employ this kind of mechanism are called "hidden variable" theories. Unfortunately, in Bohm's theory, the particles are influenced by phenomena arbitrarily far away from them. Faraday and Maxwell had argued strongly against such theories in the nineteenth century, and since that time, physicists had adopted locality — meaning that an object is influenced directly only by its immediate physical surroundings — as a basic principle of physics. For this reason, many physicists find Bohm's approach unappealing.

In 1964, inspired by Einstein, Podolsky, and Rosen's argument, John Bell, working at the European Organization for Nuclear Research (CERN), proposed an experiment to rule out any local, classical picture of the world in which influences travel no faster than the speed of light. Bell's proposal was, if you like, a way of "catching reality in the act" of behaving in a manner that would simply be impossible in any local, classical description.

The experiment Bell envisaged involved two elementary particles flying apart just as Einstein, Podolsky, and Rosen had imagined. Like them, Bell considered the two particles to be in a perfectly correlated state. However, instead of thinking of measuring their positions or velocities, Bell imagined measuring something even simpler: their spins.

Most of the elementary particles we know of have a spin — something Pauli and then Dirac had explained. You can think of particles, roughly speaking, as tiny little tops spinning at some fixed rate. The spin is quantized in units given by Planck's constant, but the details of that will not matter here. All that concerns us in this case is that the outcome is binary. Whenever you measure a particle's spin, there are only two possible outcomes: you will either find the particle spinning about the measurement axis either anticlockwise or clockwise at a fixed rate. If the particle spins anticlockwise, we say its spin is "up," and if it is clockwise, we say its spin is "down."

Bell hypothesized a situation in which the two Einstein–Podolsky–Rosen particles are produced in what is known as a "spin zero state." In such a state, if you measure both particles with respect to the same axis, then if you find one of them to be "up," the other will be "down," and vice versa. We say that the particles are "entangled," meaning that measuring the state of one fixes the state of the other. According to quantum theory, the two particles can retain this connection no matter how far apart they fly. The strange aspect of it is that by measuring one, you instantly determine the state of the other, no matter how distant it is. This is an example of what Einstein called "spooky non-locality" in quantum physics.

Bell imagined an experiment in which the particles were allowed to fly far apart before their spins were measured. He discovered a subtle but crucial effect, which meant that no local, classical picture of the world could possibly explain the pattern of probabilities that quantum theory predicts.

To make things more familiar, let us pretend that instead of two particles, we have two boxes, each with a coin inside it. Instead of saying the particle's spin is "up," we'll say the coin shows heads; and instead of saying the particle's spin is "down," we'll say the coin shows tails.

Imagine you are given two identical boxes, each in the shape of a tetrahedron — a pyramid with four equilateral triangular sides. One side is a shiny metal base, and the other three are red, green, and blue. The coloured sides are actually small doors. Each of them can be opened to look inside the pyramid. Whenever you open a door, you see a coin lying inside on the base, showing either heads or tails.

Upon playing with the boxes, you notice that the bases are magnetic and stick together base to base. When the boxes are stuck together like this, the doors are held tightly shut, and there is a soft hum indicating the state of the boxes is being set.

Now you and a friend pull the two boxes apart. This is the analogue of the Einstein–Podolsky–Rosen experiment. You each take a box and open one of its doors. First, you both look through the red door of your box. You see heads and your friend sees tails. So you repeat the experiment. You put the boxes together, pull them apart, and each of you opens the red door. After doing this many times, you conclude that each result is entirely random — half the time your coin shows heads, and half the time it shows tails. But whatever you get, your friend gets *exactly* the opposite. You try taking the boxes very far apart before opening them, and the same thing happens. You cannot predict your own result, but whatever that result turns out to be, it allows you to predict your partner's finding with certainty. Somehow, even though each box gives an apparently random result, the two boxes always give opposite results.

It's strange, but so far there is no real contradiction with a local, classical picture of the world. You could imagine that there is a little machine that makes a random choice of how to program the two boxes when they are placed base to base. If it programs the first box to show heads when the red door is opened, it will program the second box to show tails. And vice versa. This programming trick will happily reproduce everything you have seen so far.

Now you go further with the experiment. You decide that you will both open only the green door. And you find the same thing as before — each of you gets heads or tails half the time, and each result is always the opposite of the other. The same happens with the blue door.

Still, there is no real contradiction with a classical picture of the world. All that is required to reproduce what you have seen is that when the two bases are held together, one box is programmed randomly and the other box is given exactly the opposite program. For example, if the first box is programmed to give heads/ heads/tails when you open the red, green, or blue door, then the other is programmed tails/tails/heads when you open the red, green, or blue door. If the first box is programmed heads/heads/heads, the second is programmed tails/tails/tails. And so on. This arrangement would explain everything you have seen so far.

Now you try something different. Again, you put the two bases together and pull the boxes apart. But now, each of you chooses *at random* which door to open — either red, green, or blue — and records the result. Doing this again and again, many times, you find that half the time you agree and half the time you disagree. Initially, it seems like sanity has been restored: the boxes are each giving a random result. But wait! Comparing your results more carefully, you see that whenever you and your partner happen to open the same-coloured door, you always disagree. So there is still a strong connection between the two boxes, and their results are not really independent at all. The question is: could the boxes possibly have been programmed to always disagree when you open the same-coloured door but to disagree only half the time when you each open a door randomly?

Imagine, for example, that the boxes were programmed to give, say, heads/heads/tails for your box and tails/tails/heads for your friend's. You pick one of three doors, at random, and your friend does the same. So there are nine possible ways for the two of you to make your choices: red–red, red–green, red–blue, green–red, green–green, green–blue, blue–red, blue–green, and blue–blue. In five of them you will get the opposite results, with one seeing heads and the other tails, but in four you will agree. What about if the boxes were programmed heads/heads/heads and tails/tails/tails? Well, then you would always disagree. Since every other program looks like one of these two cases, we can safely conclude that *however* the boxes are programmed, if you open the doors randomly there is always at least a five-ninths chance of your disagreeing on the result. But that isn't what you found in the experiment: you disagreed half the time.

As you may have already guessed, quantum theory predicts exactly what you found. You agree half the time and disagree half the time. The actual experiment is to take two widely separated Einstein–Podolsky–Rosen particles in a spin zero state and measure their spins along one of three axes, separated by 120 degrees. The axis you choose is just like the door you pick in the pyramidal box. Quantum theory predicts that when you pick the same measurement axis for the two particles, their spins always disagree. Whereas if you pick different axes, they agree three-quarters of the time and disagree one-quarter of the time. And if you pick axes randomly, you agree half the time and disagree half the time. As we have just argued with the boxes, such a result is impossible in a local, classical theory.[21]

Before drawing this conclusion, you might worry that the particles might somehow communicate with each other, for example by sending a signal at the speed of light. So that, for example, if you chose different measurement axes, the particles would correlate their spins so that they agreed three-quarters of the time and disagreed one-quarter of the time, just as predicted by quantum mechanics. Experimentally, you can eliminate this possibility by ensuring that at the moment you choose the measurement axis, the particles are so far apart that no signal could have travelled between them, even at the speed of light, in time to influence the result.

In 1982, the French physicists Alain Aspect, Philippe Grangier, and Gérard Roger conducted experiments in which the setting for the measurement axis of Einstein–Podolsky–Rosen particles was chosen while the particles were in flight. This was done in such a way as to exclude any possible communication between the measured particles regarding this choice. Their results confirmed quantum theory's prediction, showing that the world works in ways we cannot possibly explain using classical notions. Some physicists were moved to call this physics' greatest-ever discovery.

Although the difference between five-ninths and one-half may sound like small change, it is a little like doing a very long sum and finding that you have proven that 1,000 equals 1,001 (I am sure this has happened to all of us many times, while doing our taxes!). Imagine you checked and checked again, and could not find any mistake. And then everyone checked, and the world's best computers checked, and everyone agreed with the result. Well, then by subtracting 1,000, you would have proven that 0 equals 1. And with that, you can prove any equation to be right and any equation to be wrong. So all of mathematics would disappear in a puff of smoke. Bell's argument, and its experimental verification, caused all possible classical, local descriptions of the world similarly to disappear.

These results were a wake-up call, emphasizing that the quantum world is qualitatively different from any classical world. It caused people to think carefully about how we might utilize these differences in the future. In Chapter Five, I will describe how the quantum world allows us to do things that would be impossible in a classical world. It is opening up a whole new world of opportunity ahead of us — of quantum computers, communication, and, perhaps, perception — whose capabilities will dwarf what we have now. As those new technologies come on stream, they may enable a more advanced form of life capable of comprehending and picturing the functioning of the universe in ways we cannot. Our quantum future is awesome, and we are fortunate to be living at the time of its inception.

. . .

OVER THE COURSE OF the twentieth century, in spite of Einstein's qualms, quantum theory went from one triumph to the next. Curie's radioactivity was understood to be due to quantum tunnelling: a particle trapped inside an atomic nucleus is occasionally allowed to jump out of it, thanks to the spreading out in space of its probability wave. As the field of nuclear physics was developed, it was understood how nuclear fusion powers the sun, and nuclear energy became accessible on Earth. Particle physics and the physics of solids, liquids, and gases were all built on the back of quantum theory. Quantum physics forms the foundation of chemistry, explaining how molecules are held together. It describes how real solids and materials behave and how electricity is conducted through them. It explains superconductivity, the condensation of new states of matter, and a host of other extraordinary phenomena. It enabled the development of transistors, integrated circuits, lasers, LEDs, digital cameras, and all the modern gadgetry that surrounds us.

Quantum theory also led to rapid progress in fundamental physics. Paul Dirac combined Einstein's theory of relativity with quantum mechanics into a relativistic equation for the electron, in the process predicting the electron's antiparticle, the positron. Then he and others worked out how to describe electrons interacting with Maxwell's electromagnetic fields — a framework known as quantum electrodynamics, or QED. The U.S. physicists Richard Feynman and Julian Schwinger and the Japanese physicist Sin-Itiro Tomonaga used QED to calculate the basic properties and interactions of elementary particles, making predictions whose accuracy eventually exceeded one part in a trillion.

Following a suggestion from Paul Dirac, Feynman also developed a way of describing quantum theory that connected it directly to Hamilton's action formalism. What Feynman showed was that the evolution in time of Schrödinger's wavefunction could be written using only Euler's e, the imaginary number i, Planck's constant h, and Hamilton's action principle. According to Feynman's formulation of quantum theory, the world follows all possible histories at once, but some are more likely than others. Feynman's description gives a particularly nice account of the "double-slit" experiment: it says that the particle or photon follows *both* paths to the screen. You add up the effect of the two paths to get the Schrödinger wavefunction, and it is the interference between the two paths that creates the pattern of probability for the arrival of particles or photons at various points on the screen. Feynman's wonderful formulation of quantum theory is the language I shall use in Chapter Four to describe the unification of all known physics.

As strange as it is, quantum theory has become the most successful, powerful, and accurately tested scientific theory of all time. Although its rules would never have been discovered without many clues from experiment, quantum theory represents a triumph of abstract, mathematical reasoning. In this chapter, we have seen the magical power of such thinking to extend our intuition well beyond anything we can picture. I emphasized the role of the imaginary number i, the square root of minus one, which revolutionized algebra, connected it to geometry, and then enabled people to construct quantum theory. To a large extent, the entry of i is emblematic of the way in which quantum theory works. Before we observe it, the world is in an abstract, nebulous, undecided state. It follows beautiful mathematical laws but cannot be described in everyday language. According to quantum theory, the very act of our observing the world forces it into terms we can relate to, describable with ordinary numbers.

In fact, the power of i runs deeper, and it is profoundly related to the notion of time. In the next chapter, I will describe how Einstein's theory of special relativity unified time with space into a whole called "spacetime." The German mathematician Hermann Minkowski clarified this picture, and also noticed that if he started with four dimensions of space, instead of three, and treated one of the four space coordinates as an imaginary number — an ordinary number times i — then this imaginary space dimension could be reinterpreted as time. Minkowski found that in this way, he could recover all the results of Einstein's special relativity, but much more simply.[22]

It is a remarkable fact that this very same mathematical trick, of starting with four space dimensions and treating one of them as imaginary, not only explains all of special relativity, it also, in a very precise sense, explains all of quantum physics! Imagine a classical world with four space dimensions and no time. Imagine that this world is in thermal equilibrium, with its temperature given by Planck's constant. It turns out that if we calculate all the properties of this world, how all quantities correlate with each other, and then we perform Minkowski's trick, we reproduce all of quantum theory's predictions. This technique, of representing time as another dimension of space, is extremely useful. For example, it is the method used to calculate the mass of nuclear particles — like the proton and the neutron — on a computer, in theoretical studies of the strong nuclear force.

Similar ideas, of treating time as an imaginary dimension of space, are also our best clue as to how the universe behaves in black holes or near the big bang singularity. They underlie our understanding of the quantum vacuum, and how it is filled with quantum fluctuations in every field. The vacuum energy is already taking over the cosmos and will control its far future. So, the imaginary number i lies at the centre of our current best efforts to face up to the greatest puzzles in cosmology. Perhaps, just as i played a key role in the founding of quantum physics, it may once again guide us to a new physical picture of the universe in the twenty-first century.

Mathematics is our "third eye," allowing us to see and understand how things work in realms so remote from our experience that they cannot be visualized. Mathematicians are often viewed as unworldly, working in a dreamed-up, artificial setting. But quantum physics teaches us that, in a very real sense, we all live in an imaginary reality.

THREE

WHAT BANGED?

"The known is finite, the unknown infinite; intellec-
tually we stand on an islet in the midst of an
illimitable ocean of inexplicability. Our business in
every generation is to reclaim a little more land."
— T. H. Huxley, 1887[1]

"Behind it all is surely an idea so simple, so beautiful,
that when we grasp it — in a decade, a century, or a
millennium — we will all say to each other, how could it
have been otherwise?" — John Archibald Wheeler, 1986[2]

SOMETIMES I THINK I'M the luckiest person alive. Because
I get to spend my time wondering about the universe.
Where did it come from? Where is it going? How does it
really work?

In 1996, I took up the Chair of Mathematical Physics
at the University of Cambridge. It was an opportunity for

me to meet and to work with Stephen Hawking, holder of the Lucasian Chair — the chair Isaac Newton held. It was Hawking who, three decades earlier, had proved that Einstein's equations implied a singularity at the big bang — meaning that all the laws of physics fail irretrievably at the beginning of the universe. In the eighties, along with U.S. physicist James Hartle, Hawking had also proposed a way to avoid the singularity so that the laws of physics could describe how the universe began.

During the time Stephen and I were working together, he invited me to be interviewed on TV as part of a series about the cosmos he was helping to produce. Soon after the program was aired, a letter appeared in my mailbox. It was from Miss Margaret Carnie, my grade school teacher in Tanzania. I jumped up and down with joy. Margaret had always held a special place in my heart, but we had lost touch when I was ten years old and my family moved to England. Margaret was now back in Scotland and had spotted my name on TV. She wrote: "Are you the same Neil Turok who I taught as a little boy at Bunge Primary School in Dar es Salaam?"

Margaret had devoted her life to teaching. She was a part of a long tradition of maths and science teaching dating back to the Scottish Enlightenment. She and her identical twin sister, Ann, had both taught in the little government school in Dar es Salaam and lived with their mother, also a teacher, in the flat above. The secret of Margaret's approach was not to instruct her students but to gently point them in interesting directions. She gave me lots of freedom, and lots of materials. I made plans and maps of the school, drew living creatures and plants, experimented with Archimedes' principle, played with trigonometry, and explored mathematical formulae. At home, I made electric motors and dynamos from old car parts, collected beetles, watched ant lions for hours, caused explosions, and made huts out of palm fronds — though you had to watch out for the snakes moving in! It was a wonderful childhood.

Just before Margaret had written to me, I'd been told about one of Cambridge's oldest traditions, that newly appointed professors were invited to give an inaugural public lecture. I'd also learned that very few bothered anymore. Having heard from Margaret, I simply had to give mine in her honour. So I called her up and invited her and her sister, and after that we kept in touch regularly. A few years later, I visited them in Edinburgh. They were in their late seventies but still very active — volunteering for the museum, organizing and attending public lectures, and generally being the life and soul of their community.

As we sat in their little apartment drinking tea, Margaret asked me what I was working on. When I answered "Cosmology" and started to explain, she waved all the details away. She said in her strong Scottish accent, "There's only one really important question. Every time I go to a public lecture on astronomy, I ask the lecturer: 'What banged?' But I never get a sensible reply."

"Margaret, that's exactly what I'm working on!" I said. "I always knew you were a clever boy," she replied. And she pulled out an old photo of me grinning at her while wielding a hoe in the schoolyard farm. I did look pretty enthusiastic.

I tried to explain to Margaret the latest model I'd been working on, where the big bang is a collision between two three-dimensional worlds. But her eyes glazed over and I could tell that she thought it was all too complicated and technical. She was a down-to-earth, pragmatic person. She wanted a simple, straightforward, and believable answer. Sadly, she and her sister passed away a few years ago. I'm still looking for an answer that would satisfy them.

IN THIS MODERN AGE, where our lives are so focused on human concerns, cosmology may seem like a strange thing to be thinking about. Einstein to some extent was expressing this when he said, "The most incomprehensible thing about the world is that it is comprehensible."[3] Even he thought it a surprise that *we* — people — can look out at *it* — the universe — and understand how it works.

In ancient Greece, they saw things quite differently. The early Greek philosophers viewed themselves as a part of the natural world, and for them understanding it was basic to all other endeavours. They thought of the universe as a divine, living being whose innermost essence was harmony. They called it *kosmos*, and believed it to represent the ultimate truth, wisdom, and beauty. The word "theory," or *theoria*, which the ancient Greeks also invented, means "I see (*orao*) the divine (*theion*)."[4] They believed nature's deep principles should guide our notions of justice and of how to best organize society. The universe is far greater than any of us, and through its contemplation we may gain a proper perspective of ourselves and how we should live. According to the ancient Greeks, understanding the universe is not a surprise: it is the key to who we are and should be. They believed the universe to be comprehensible, and history has certainly proven them right. We should draw encouragement from their example and gain optimism and belief in ourselves.

In the Middle Ages, cosmology also played a major role in society. There was a great debate pitting Renaissance thinkers against the Catholic Church over its insistence that the Earth was the centre of the universe. In overturning this view, Copernicus and Galileo revived many of the ideals of ancient Greece, including the power of rational argument over dogma. In showing that Earth is just a planet moving around the sun, they liberated us from the centre of the cosmos: we are space travellers, with a whole universe to explore. Galileo's proposal, inspired by the ancient Greeks, of universal, mathematical laws was a profoundly democratic idea. The world could be understood by anyone, and the only tools one needed were reason, observation, and mathematics, none of which depended on your position or authority.

Built upon Galileo's intuition, Isaac Newton's unification of the laws of motion and gravity, along with
his invention of calculus, laid the foundations for all of
engineering and the industrial age. But remember that
Newton's key discoveries were learned from the solar
system, though all of their applications for the next few
centuries were terrestrial. The universe has been very
good at teaching us things.

Two hundred years later, Michael Faraday learned
more secrets from nature. His experiments and his
intuition guided Maxwell, just as Galileo's had guided
Newton. Maxwell unified electricity, magnetism, and
light, and laid the ground for quantum theory and relativity. Einstein was so impressed with Maxwell's theory
that, in his 1905 paper on the quantization of light, he
wrote, "The wave theory of light...has proven to be an
excellent model of purely optical phenomena and presumably will never be replaced by another theory."[5] And
later, in an essay on Maxwell in 1931, Einstein wrote,
"Before Maxwell, physical reality was thought of as
consisting of material particles...Since Maxwell's time,
physical reality has been thought of as represented by
continuous fields...This change in the conception of
reality is the most profound and most fruitful that physics has experienced since the time of Newton."[6]

Maxwell's theory inspired Einstein's theory of space,
time, and gravity, which would eventually describe the
whole cosmos. Its quantum version would lead to the
development of subatomic physics and the description of

the hot big bang. And late in the twentieth century, the quantum theory of fields would produce theories explaining the density variations that gave rise to galaxies. Physics has come full circle: what Newton gleaned from the heavens inspired the development of mathematical and physical theories on Earth. Now these in turn have extended our understanding of the cosmos. Think of how delighted the ancient Greek thinkers would have been with this progress. The virtuous cycle of learning from the universe and extending the reach of our knowledge continues, and we need to take heart from it.

When I was a very young child, as I have already mentioned, I thought the sky above was a vaulted ceiling with painted-on stars. But when I stand and look out at the universe now, I appreciate the extent to which we have been able to comprehend it through the work of Maxwell, Einstein, and all the others who followed. To me, this understanding is an incredible privilege, and a glimpse of what we are and should be. When we look up at the sky, we're actually seeing inside ourselves. It is an act of wonder to stand there and realize how the world really works. And even more so to peer over the edge of our understanding, and anticipate the answers to even bigger questions. The mathematics that gets us there is a means to an end. For me, the real world is the awesome thing, and what I'm most interested in is what it all means.

· · ·

IMAGINE A PERFECT BALL of light, just a millimetre across. It is the brightest, most intense light you can possibly conceive of. If you can think of compressing the sun down to the size of an atomic nucleus, that will give you some sense of the searing brilliance inside the ball. At these extreme temperatures, it is far too hot for any atoms or even atomic nuclei to survive. Everything is broken down into a plasma of elementary particles and photons, the energy packets of light.

Now imagine the ball of light expanding, faster than anything you have ever seen or can imagine. Within one second, it is a thousand light years across. It didn't get there by the light and particles travelling outwards in an explosion — nothing can travel that fast. Instead, *the space inside the ball expanded.* As it grew, the wavelength of the photons was stretched out. They became far less energetic, and the plasma temperature fell. One second after the expansion began, the temperature is ten billion degrees. The photons are still energetic enough to tear atomic nuclei apart.

As the space within the ball expands and the plasma cools further, the matter particles are able to clump into atomic nuclei. Ten minutes after the expansion began, the atomic nuclei of the lightest chemical elements — hydrogen, helium, lithium — are formed. The nuclei of the heavier chemical elements, such as carbon, nitrogen, and oxygen, will form later in stars and supernovae.

The ball of light continues to expand, at an ever decreasing rate. After four hundred thousand years, it is ten million light years across. The conditions are cool enough now for the atomic nuclei to gather electrons around them and form the first atoms. The conditions are similar to those at the surface of the sun, with the temperature measured in thousands of degrees. Space is still expanding, though at a far slower rate, and it is still filled with an almost perfectly uniform plasma consisting of matter and radiation. However, as we look across space, we see small variations in the density and temperature from place to place, at a level of just one part in a hundred thousand. These mild ripples in the density occur on all scales, small and large, like a pattern of waves on the ocean.

As the universe expands, gravity causes the ripples to grow in strength, like waves approaching the shore. The slightly denser regions become much more dense and collapse like giant breakers to form galaxies, stars, and planets. The slightly less dense regions expand out into the empty voids between the galaxies. Today, 13.7 billion years after the expansion began, the millimetre-sized brilliant ball of light has grown to a vast region encompassing hundreds of billions of galaxies, each containing hundreds of billions of stars.

Although the events I have just described were in our past, we can check that they happened just by looking out into space. Since light travels at a fixed speed, the farther out we look, the younger the objects we witness. The moon, for example, is a light second away, meaning that we see it as it was a second ago. Likewise we see the sun as it was eight minutes ago, and Jupiter as it was forty minutes ago. The nearest stars are ten light years beyond the solar system, and we see them as they were a decade ago. The Andromeda Galaxy, one of our nearest galaxy neighbours, is two million light years away, so we are seeing it as it was before our species appeared on Earth.

As we look farther out, we see farther back in time. Around us is the middle-aged universe: quieter and more predictable, slowly spreading. By detecting chemical elements within stars, we can measure their abundances throughout the universe and check that they agree with predictions. Reaching out to around twelve billion light years, we see the universe's tumultuous adolescence with the collapsing clouds of matter creating quasars — powerful sources of radiation formed around massive black holes, as well as newly formed spiral and elliptical galaxies. Beyond those, we see baby galaxies, some just nascent wisps of gas starting to pull themselves together. Looking out farther, we can see all the way back to a time just four hundred thousand years after the big bang, when space was filled with a hot plasma at a temperature roughly that of the surface of the sun.

We can see no farther, because at earlier times the atoms were broken up into charged particles, which scatter light and obscure our view of the earlier universe. The radiation that emanates from this hot plasma rind all around us has been stretched out by the expansion of the universe to microwave wavelengths. As we look back

FROM QUANTUM TO COSMOS

to this epoch, it appears from our perspective as if we are in the centre of a giant, hot, spherical microwave oven.

We have just described the hot big bang theory, a spectacularly successful description of the evolution of the universe. "But what banged?" I hear you ask. There is no bang in the picture, just the expansion of space from a very dense assumed starting point. Space expanded in the same way and at once, everywhere. There was no centre of the expansion: the conditions in the universe were the same across all of space. Our millimetre-sized ball is just the portion of the primeval universe that expanded into everything we can now see today.

IN 1982, I WAS a graduate student at Imperial College, London, and just beginning to get interested in cosmology. I heard about a workshop taking place at Cambridge called "The Very Early Universe," and I went there for a day to listen to the talks. All the most famous theorists were there: Alan Guth, Stephen Hawking, Paul Steinhardt, Andrei Linde, Michael Turner, Frank Wilczek, and many others. And they were all very excited about the theory of inflation.

The goal of inflation was to explain the initial ball of light. The ball had many puzzling features. In addition to being extremely dense, it must have been extremely smooth throughout its interior. The space within it was not curved, as Einstein's theory of gravity allowed, but almost perfectly flat. How could it be that such an object emerged at the beginning of the universe? And how could it have produced the tiny density variations needed to seed the formation of galaxies?

The theory of inflation was invented by MIT physicist Alan Guth as a possible explanation. Guth's idea was that even if the very early universe was random and chaotic, there might be a mechanism for smoothing it out and filling it with a vast amount of radiation. He thought he had found such a mechanism in grand unified theories, which attempted to connect our description of all the particles in nature and all the forces except gravity. In these theories, there are certain kinds of fields, called "scalar fields," which take a value at each point in space. They are similar to electric and magnetic fields, but even simpler in that they have only a value, not a direction, at each point. In grand unified theories, sets of these scalar fields, called "Higgs fields," were introduced in order to distinguish between the different kinds of particles and forces. They were generalizations of the electroweak Higgs field, which we shall discuss in Chapter Four, recently reported to have been discovered at the Large Hadron Collider.

These theories postulated a form of energy called "scalar potential energy," which unlike ordinary matter was gravitationally repulsive. Guth imagined a tiny patch of the universe starting out full of nothing but this energy. Like our ball of light, it would be extremely dense. Its repulsive gravity would accelerate the expansion of space even faster than the interior of our ball of light, causing space to grow exponentially in its first phase. Guth called this scenario "cosmic inflation."

In Guth's picture the universe might have started from a region far smaller than a millimetre, far smaller even than an atomic nucleus, and containing far less energy. In fact, you could contemplate the universe starting out with a patch of space not much larger than the Planck length, a scale believed to be an ultimate limit imposed by quantum theory. And it need contain only as much energy as the chemical energy stored in an automobile's gas tank.[7] The inflationary expansion of space, filled with scalar potential energy at a fixed density, would create all the energy in the universe from a tiny seed. Guth called this effect the "ultimate free lunch." The notion is beguiling but, as I shall discuss later, potentially misleading because energy is not constant when space expands. The idea that you might get "something for nothing" nevertheless underlies much of inflationary thinking. Upon more careful examination, as we shall see, there is always a price to pay.

If a tiny patch of the universe started out in this state, the scalar potential energy would blow it up exponentially, almost instantly making it very large, very uniform, and very flat. When it reached a millimetre in size, you could imagine the scalar potential energy decaying into radiation and particles, producing a region like the ball of light at the start of the big bang. In Guth's picture, the scalar potential energy was a sort of self-replicating dynamite. Just a tiny piece of it would be enough to create the initial conditions for the hot big bang.

Inflation brought an unexpected bonus: a quantum mechanism for producing the small density variations —

the cosmic ripples — that later seeded the formation of galaxies. The mechanism is based on quantum mechanics: the scalar potential energy develops random variations as a consequence of Heisenberg's uncertainty principle, causing it to vary from place to place, on microscopic scales. The exponential expansion of the universe blows up these tiny ripples into very large-scale waves in the density of the universe. These density waves are produced on all scales, and it was a triumph for inflation that the density waves were predicted to have roughly the same strength on every scale. The level of the density variation in these waves can be adjusted by a careful tuning of the inflationary model to one part in a hundred thousand, the level of density variations required to explain the origin of galaxies.

As a young scientist, I was amazed to see the confidence that these theorists placed in their little equations when describing a realm so entirely remote from human experience. There was no direct evidence for *anything* they were discussing: the exponential blow-up of the universe during inflation, the scalar fields and their potential energy which they hoped would drive it, and — what they were most excited about at the meeting — the vacuum quantum fluctuations that they hoped inflation would stretch and amplify into the seeds of galaxies. Of course, they drew their confidence from physics' many previous successes in explaining how the universe worked with mathematical ideas and reasoning.

But there seemed to me a big difference. Maxwell and Einstein and their successors had been guided by a profound belief that nature works in simple and elegant ways. Their theories had been extremely conservative, in the sense of introducing little or no arbitrariness in their new physical laws. Getting inflation to work was far more problematic. The connection to grand unified theory sounded promising, but the Higgs fields, which were introduced in order to separate the different particles and forces, would typically not support the kind of inflation needed: they would either hold the universe stuck in an exponential blow-up forever or they would end the inflation too fast, leaving the universe curved and lumpy. Working models of inflation required a fine tuning of their parameters and strong assumptions about the initial conditions. Inflationary models looked to me more like contrivances than fundamental explanations of nature.

At the same time, the attention theorists were now giving to cosmology was enormously energizing to the field. Although the inflationary models were artificial, their predictions gave observers a definite target to aim at. Over the next three decades, the inflationary proposal, along with other ideas linking fundamental physics to cosmology, helped drive a vast expansion of observational efforts directed at the biggest and most basic questions about the universe.

THE STORY OF MODERN cosmology begins with Einstein's unification of space, time, energy, and gravity, which closely echoed Maxwell's unification of electricity, magnetism, and light. When Einstein visited London, a journalist asked him if he had stood on the shoulders of Newton. Einstein replied, "That statement is not quite right; I stood on Maxwell's shoulders."[8] Just as happened with Maxwell's theory, many spectacular predictions would follow from Einstein's. Maxwell's equations had anticipated radio waves, microwaves, X-rays, gamma rays — the full spectrum of electromagnetic radiation. Einstein's equations were even richer, describing not only the fine details of the solar system but everything from black holes and gravitational waves to the expansion and evolution of the cosmos. His discoveries brought in their wake an entirely new conception of the universe as a dynamic arena. Einstein's theory was more complicated than Maxwell's, and it would take time to see all of its implications.

The most spectacular outcome of Maxwell's unified theory of electricity and magnetism had been its prediction of the speed of light. This prediction raised a paradox so deep and far-reaching in its implications that it took physicists decades to resolve. The paradox may be summarized in the simplest of questions: the speed of light relative to what? According to Newton, and to everyday intuition, if you see something moving away and chase after it, it will recede more slowly. If you move fast enough, you can catch up with it or overtake it. An absolute speed is meaningless.

In every argument, there are hidden assumptions. The more deeply they are buried, the longer it takes to reveal them. Newton had assumed that time is absolute: all observers could synchronize their clocks and, no matter how they moved around, their clocks would always agree. Newton had also assumed an absolute notion of space. Different observers might occupy different positions and move at different velocities, but again they would always agree on the relative positions of objects and the distances between them.

It took Einstein to realize that these two seemingly reasonable assumptions — of absolute time and space — were incompatible with Maxwell's theory of light. The only way to ensure that everyone would agree on the speed of light was to have them each experience *different* versions of space and time. This does not mean that the measurements of space and time are arbitrary. On the contrary, there are definite relations between the measurements made by different observers.

The relations between the measurements of space and time made by different observers are known as "Lorentz transformations," after the Dutch physicist Hendrik Lorentz, who inferred them from Maxwell's theory. In creating his theory of relativity, Einstein translated Lorentz's discovery into physical terms, showing that Lorentz's transformations take you from the positions and times measured by one observer to those measured by another. For example, the time between the ticks of a clock or the distance between the ends of a ruler depends on who makes the observation. For an observer moving past them, a clock goes more slowly and a ruler aligned with the observer's motion appears shorter than for someone who sees them at rest. These phenomena are known as "time dilation" and "Lorentz contraction," and they become extremely important when observers move relative to one another at speeds close to the speed of light.

The Lorentz transformations mix up the space and time coordinates. Such a mixing is impossible in Newton's theory, because space and time are entirely different quantities. One is measured in metres, the other in seconds. But once you have a fundamental speed, the speed of light, you can measure both times and distances in the same units: seconds and light seconds, for example. This makes it possible for space and time to mix under transformations. And because of this mixing, they can be viewed as describing a single fundamental entity, called "spacetime."

The unification of space and time in Einstein's theory, which he called "special relativity," allowed him to infer relationships between quantities which, according to Newton, were not related. One of these relations became the most famous equation in physics.

IN 1905, THE SAME year that he introduced his theory of special relativity, Einstein wrote an astonishing little three-page paper that had no references and a modest-sounding title: "Does the Inertia of a Body Depend Upon Its Energy-Content?" This paper announced Einstein's iconic formula, $E = mc^2$.

Einstein's formula related three things: energy, mass, and the speed of light. Until Einstein, these quantities were believed to be utterly distinct.

Energy, at the time, was the most abstract of them: you cannot point at something and say, "That is energy," because energy does not exist as a physical object. All you can say is that an object *possesses* energy. Nevertheless, energy is a very powerful idea, because under normal circumstances (not involving the expansion of space), while it can be converted from one form into another, it is never created or destroyed. In technical parlance, we say energy is *conserved*.

The concept of mass first arose in Newton's theory of forces and motion, as a measure of an object's inertia: how much push is required to accelerate the object. Newton's second law of motion tells you the force you need to exert to create a certain acceleration: force equals mass times acceleration.

So how does energy equal the mass of an object times the speed of light squared? Einstein's argument was simple. Light carries energy. And objects like atoms or molecules can absorb and emit light. So Einstein just looked at the process of light emission from an atom, from the points of view of two different observers.

The first observer sees the atom at rest emit a burst of electromagnetic waves. From energy conservation, it follows that the atom must have had more energy before it emitted the light than it had afterward. Now let's look at the same situation from the point of view of a second observer, moving relative to the first. The second observer sees the atom moving, both before and after the emission. According to the second observer, the atom has some energy of motion, or kinetic energy. The second observer also sees a slightly more energetic burst of radiation compared to the first, just because she is in motion. This extra energy can be calculated from Maxwell's theory, using a Lorentz transformation.

Now Einstein just wrote down the equations for energy conservation. The total energy before the emission must equal the energy after it, according to both observers. From these two equations it follows that the atom's kinetic energy after the emission, as seen by the second observer, must equal the atom's kinetic energy before the emission plus the extra energy in the burst of radiation. This equation relates the energy in the burst of radiation to the mass of the atom before and after the emission. And the equation implies that the atom's mass changes by the energy it emits divided by the square of the speed of light. If the atom loses *all* of its mass in this process, and just decays completely into the burst of radiation, the same relation applies. The amount of radiation energy released must be equal to the original mass times the speed of light squared, or $E = mc^2$.

Einstein put it this way: "Classical physics introduced two substances: matter and energy. The first had weight, but the second was weightless. In classical physics we had two conservation laws: one for matter, the other for energy. We have already asked whether modern physics still holds this view of two substances and the two conservation laws. The answer is: No. According to the theory of relativity, there is no essential distinction between mass and energy. Energy has mass and mass represents energy. Instead of two conservation laws we have only one, that of mass-energy."[9] $E = mc^2$ is a unification. It tells us that mass and energy are two facets of the same thing.

What Einstein's magical little formula tells us is that we are *surrounded* by vast stores of energy. For example, that sachet of sugar you are about to stir into your coffee has a mass energy equivalent to a hundred kilotons of TNT — enough to level New York. And of course, his discovery prefigured the development of nuclear physics, which eventually led to nuclear energy and the nuclear bomb.

In Newton's theory, there was no limit to the speed of an object. But in Einstein's theory, nothing travels faster than light. The reason is fundamental: if something *did* travel faster than light, then according to Lorentz's transformations, some observers would see it going backward in time. And that would create all sorts of causality paradoxes.

IN DEVELOPING THE THEORY of relativity, the next question facing Einstein, which echoed concerns raised by Michael Faraday more than half a century earlier, was whether the force of gravity could really travel faster than light. According to Newton, the gravitational force of attraction exerted by one mass on any other mass acts *instantaneously* — that is, it is felt immediately, right across the universe. As a concrete example, the tides in Earth's oceans are caused by the gravitational attraction of the moon. As the moon orbits Earth, the masses of water in the oceans follow. According to Newton, the moon's gravity is felt instantly. But moonlight takes just over a second to travel from the moon to Earth. Faraday and Einstein both felt it unlikely that the influence of gravity travelled any faster.

In constructing a theory of gravity consistent with relativity, one of the key clues guiding Einstein was something that Galileo had noticed: all objects fall in the same way under gravity, whatever their mass. An object in free fall behaves as if there is no gravity, as we know from the weightlessness that astronauts experience in space: an astronaut and her space capsule fall together. This behaviour suggested to Einstein that gravity was not the property of an object, but was instead a property of spacetime.

What then is gravity? Gravity is replaced, in Einstein's theory, by the bending of space and time caused by matter. Earth, for example, distorts the space-time around it, like a bowling ball sitting in the centre of a trampoline. If you roll marbles inwards, the curved surface of the trampoline will cause them to orbit the bowling ball, just as the moon orbits Earth. As the physicist John Wheeler would later put it, "Matter tells spacetime how to curve, and spacetime tells matter how to move."[10]

After ten years of trying, in 1916 Einstein finally discovered his famous equation — now called Einstein's equation — according to which the curvature of spacetime is determined by the matter contained within it. He used the mathematical description of curved space invented by the German mathematician Bernhard Riemann in the 1850s. Before Riemann, a curved surface, such as a sphere, had always been thought of as embedded within higher dimensions. But Riemann showed how to define the key concepts in geometry, like straight lines and angles, intrinsically within the curved surface, without referring to anything outside it. This discovery was very important, because it allowed one to imagine that the universe was curved, without it having to be embedded inside anything else.

Einstein's new theory, which he called "general relativity," brought our view of the universe much closer to that of the ancient Greeks: the universe as a vital, dynamic entity with a delicate balance between its elements — space, time, and matter. Einstein altered our view of the cosmos, from the inert stage I had envisaged as a child to a changeable arena that could curve or expand.

In welcoming Einstein to London, the celebrated playwright George Bernard Shaw told a jokey story about how a young professor — Albert Einstein — had demolished the Newtonian picture of the world. Upon learning that Newton's gravity was no more, people asked him: "But what about the straight line? If there is no gravitation, why do not the heavenly bodies travel in a straight line right out of the universe?" And, Shaw continues, "The professor said, 'Why should they? That is not the way the world is made. The world is not a British rectilinear world. It is a curvilinear world, and the heavenly bodies go in curves because that is the natural way for them to go.' And at last the whole Newtonian universe crumbled up and vanished, and it was succeeded by the Einsteinian universe."[11]

In the early days Max Born described Einstein's theory of general relativity thus: "The theory appeared to me then, and it still does, the greatest feat of human thinking about nature, the most amazing combination of philosophical penetration, physical intuition, and mathematical skill. But its connections with experience were slender. It appealed to me like a great work of art to be admired from a distance."[12] Today, Born's statement is no longer true. Einstein's younger successors applied his theory to the cosmos and found it to work like a charm. Today, general relativity is cosmology's workhorse, and almost every observation we make of the universe — whether from the Hubble Telescope or giant radio arrays or X-ray or microwave satellites — relies on Einstein's theory for its interpretation.

General relativity is not an easy subject. As an undergraduate, I trudged my way through a famously massive textbook on the subject called *Gravitation*, which weighs 2.5 kilograms. It was a quixotic attempt — while the subject is conceptually simple, its equations are notoriously difficult. After six months of trying, I decided instead to take a course. That made it all so much easier. Physics, like many other things, is best learned in person. Seeing someone else do it makes you feel you can do it too.

. . .

THE DISCOVERY OF GENERAL relativity, and its implication that spacetime was not rigid, raised a question: what is the universe doing on the very largest scales, and how is it affected by all the matter and energy within it? Like everyone else at the time, when Einstein started to think about cosmology, he assumed the universe was static and eternal. But immediately a paradox arose. Ordinary matter attracts other ordinary matter under gravity, and a static universe would just collapse under its own weight. So Einstein came up with a fix. He introduced another, simpler form of energy that he called the "cosmological term." Its main properties are that it is absolutely uniform in spacetime, and it looks exactly the same to any observer. The best way to visualize the cosmological term is as a kind of perfectly elastic, stretchy substance, like a giant sponge filling space. It has a "tension," or negative pressure, meaning that as you stretch it out it stores up energy just like an elastic band. But no matter how much you stretch it, its properties do not change — you just get more of it.

At first, a negative pressure sounds like exactly what you don't want holding up the universe. It would suck things inward and cause a collapse. However, as we described earlier, the expansion of the universe is not like ordinary physics. It is not an explosion: it is the expansion of space. And it turns out that the effect of a negative pressure, in the Einstein equations, is exactly the opposite of what you might expect. Its gravity is *repulsive* and causes the size of the universe to blow up. (This effect of repulsive gravity is the same one that Guth used in his theory of inflation.)

So Einstein made his mathematical model of the universe stand still by balancing the attractive gravity of the ordinary matter against the repulsive gravity of his cosmological term. The model was a failure. As noticed by the English astrophysicist Arthur Eddington, the arrangement is unstable. If the universe decreased a little in size, the density of the ordinary matter would rise and its attraction would grow, causing the universe to collapse. Likewise, if the universe grew a little in size, the matter would be diluted and the cosmological term's repulsion would win, blowing the universe up.

It would fall to two very unusual people to see what Einstein could not: that his theory describes an expanding universe.

THE FIRST WAS ALEXANDER Friedmann, a gifted young Russian mathematical physicist who had been decorated as a pilot in the First World War. Due to the war and the Russian Revolution that followed it, news of Einstein's theory of general relativity did not reach St.

Petersburg, where Friedmann worked, until around 1920. Nevertheless, within two years, Friedmann was able to publish a remarkable paper that went well beyond Einstein. Like Einstein, Friedmann assumed the universe, including ordinary matter and the cosmological term, to be uniform across space and in all directions. However, unlike Einstein, he did not assume that the universe was static. He allowed it to change in size, in accordance with Einstein's equation.

What he discovered was that Einstein's static universe was completely untypical. Most mathematical solutions to Einstein's equations described a universe which was expanding or collapsing. Einstein reacted quickly, claiming Friedmann had made mathematical errors. However, within a few months, he acknowledged that Friedmann's results were correct. But he continued to believe they were of exclusively mathematical interest, and would not match the real universe. Remember, at the time of these discussions, very little was known from observations. Astronomers were still debating whether our own Milky Way was the only galaxy in the universe, or whether the patchy clouds called "nebulae," seen outside its plane, were distant galaxies.

Einstein's reason for disliking Friedmann's evolving universe solutions was that they all had singularities. Tracing an expanding universe backward in time, or a collapsing universe forward in time, you would typically find that at some moment all of space would shrink to a point and its matter density would become infinite. All the laws of physics would fail at such an event, which we call a "cosmic singularity."

Friedmann nevertheless wondered what would happen if you followed the universe through a singularity and out the other side. For example, in some models he studied, the universe underwent cycles of expansion followed by collapse. Mathematically, Friedmann found he could continue the evolution through the singularity and out into another cycle of expansion and collapse. This idea was again prescient, as we will discuss later.

Today, Friedmann's mathematical description of the expansion of the universe provides the cornerstone of all of modern cosmology. Observations have confirmed its predictions in great detail. But Friedmann never saw his work vindicated. In the summer of 1925, he made a record-breaking ascent in a balloon, riding to 7,400 metres, higher than the highest mountain in all Russia. Not long afterwards, he became ill with typhoid and died in hospital.

TWO YEARS LATER, UNAWARE of Friedmann's work, a Belgian Jesuit, Abbé Georges Lemaître, was also considering an evolving universe. Lemaître added a new component: radiation. He noticed that the radiation would slow the expansion of the universe. He also realized that the expansion would stretch out the wavelength of electromagnetic waves travelling through space, causing the light emitted from distant stars and galaxies to redden as it travelled towards us. The U.S. astronomer Edwin Hubble had already published data showing a reddening of the starlight from distant galaxies. Lemaître interpreted Hubble's data to imply that the universe must be expanding. And if he traced the expansion back in time, billions of years into our past, he found the size of the universe reached zero: it must have started at a singularity.

Once more, Einstein resisted this conclusion. When he met Lemaître in Brussels later that year, he said, "Your calculations are correct, but your grasp of physics is abominable."[13] However, he was once more forced to retract. In 1929, Hubble's observations confirmed the reddening effect in detail and were quickly recognized as confirming Lemaître's predictions. Still, many physicists resisted. As Eddington would say, the notion of a beginning of the world was "repugnant."[14]

Lemaître continued to pursue his ideas, trying to replace the singular "beginning" of spacetime with a quantum phase. What he had in mind was to use quantum theory to prevent a singularity at the beginning of the universe, just as Bohr had quantized the orbits of electrons in atoms to prevent them from falling into the nucleus. In a 1931 article in *Nature*, Lemaître stated: "If the world has begun with a single quantum, the notions of space and time would altogether fail to have any meaning at the beginning: they would only begin to have a sensible meaning when the original quantum had been divided into a sufficient number of quanta. If this suggestion is correct, the beginning of the world happened a little before the beginning of space and time."[15]

Lemaître called his hypothesis "the Primeval Atom," and as we'll see, it prefigured the ideas of the 1980s.

In January 1933, there was a charming encounter between Einstein and Lemaître in California, where they had both travelled for a seminar series. According to an article in the *New York Times*,[16] at the end of Lemaître's talk, Einstein stood and applauded, saying, "This is the most beautiful and satisfactory explanation of creation to which I have ever listened." And they posed together for a photo, which appeared with the caption: "Einstein and Lemaître — They have a profound admiration and respect for each other."

If ever one needed confirmation of the idea that diversity feeds great science, it is provided by these strange encounters between Einstein, and, first, a Russian aviator, and then a Belgian priest. The meetings must have been intense: our whole understanding of the cosmos rested in the balance.

SHORTLY AFTER THE SECOND World War, George Gamow, a former student of Friedmann's at St. Petersburg, made the next big step forward by bringing nuclear physics into cosmology. "Geo," as he was known to his friends, was a Dionysian personality — fond of jokes and pranks, a heavy drinker, larger than life. His great strength as a scientist was his audacious insight: he wasn't one to bother with details, but enjoyed the big picture. He did much to encourage others to work on interesting problems, especially applications of nuclear physics.

For example, in 1938, he and Edward Teller, best known as "the father of the hydrogen bomb," organized a conference called "Problems of Stellar Energy Sources," bringing physicists and astronomers together to work out the nuclear processes that power the sun and other stars. This historic meeting launched the modern description of stars in the universe, one of the most successful areas of modern science.

After the Russian Revolution, Gamow had been the first student to leave on an international exchange, spending time in Copenhagen with Bohr and then in Cambridge with Rutherford. Eventually, and with many regrets, he defected to the West and became a professor at George Washington University in Washington, D.C. No doubt due to his Russian connections, Gamow was not cleared to work on the atomic bomb, but he did consult for the U.S. Navy's Bureau of Ordnance. In this capacity he played the role of go-between, carrying documents up to Princeton for Einstein to examine at his home there. Apparently, they would work on these documents in the mornings and discuss cosmology in the afternoons.[17]

Gamow was an expert in nuclear physics. In 1928, he had explained the radioactive decay of heavy atomic nuclei, discovered by Curie, as being due to the quantum tunnelling of subatomic particles out of the nuclei's interiors. When Gamow started thinking about cosmology, his goal was typically ambitious: to explain the abundances of every chemical element in nature, from hydrogen through the whole periodic table.

His idea was simplicity itself: if the early universe was hot, it would behave like a giant pressure cooker. At very high temperatures, space would be filled with a plasma of the most basic particles — like electrons, protons, and neutrons. At those temperatures, they would be flying around so fast that none of them could stick together. As the universe expanded, it would cool, and the neutrons and protons would stick together to form atomic nuclei.

His student Ralph Alpher, and another young collaborator, Robert Herman, worked out the details, combining Friedmann and Lemaître's equations with the laws of nuclear physics to figure out the abundances of the chemical elements in the universe today. The approach worked well for the lightest elements — hydrogen and its heavier isotopes, and helium and lithium — and is now the accepted explanation for their relative abundances. But it failed to explain the formation of heavier elements, like carbon, nitrogen, and oxygen, later understood to have formed in stars and supernovae, and for this reason did not attract the interest it deserved.

Lying buried in Alpher, Gamow, and Herman's papers was a most wonderful prediction: the hot radiation that filled space in the early universe would never entirely disappear. One second after the singularity, when the first atomic nuclei started to form, the radiation's temperature would have been billions of degrees. As the universe expanded, the temperature of the radiation would fall, and today it would be just a few degrees above absolute zero.

Alpher, Gamow, and Herman realized that today's universe should be awash with this relic radiation — the cosmic microwave background radiation. In terms of its total energy density, it would greatly exceed all of the energy ever radiated by every star formed in the universe. And its spectrum — how much energy there is in the radiation at each wavelength — would be the same as that of a hot object, the very same spectrum described by Max Planck in 1900 when he proposed the quantization of light.

· · ·

UP TO THIS POINT, discussions of the hot big bang theory were almost entirely theoretical. That situation was about to change dramatically. In 1964, working in Holmdel, New Jersey, Arno Penzias and Robert Wilson unintentionally detected the cosmic microwave background radiation. Their instrument was a giant, ultrasensitive radio antenna that had been built at Bell Labs, AT&T's research laboratory in New Jersey, for bouncing radio signals off a large metallic balloon in space, called Echo 1. The experiments with Echo 1 were part of the effort to develop global communications technologies following the wartime development of radar.

The trials with Echo 1 were followed by the deployment of the first global communications satellite, Telstar. Launched in 1962 and looking a bit like R2-D2, the

170-pound satellite was blasted into space on the back of a modified missile. Telstar relayed the first transcontinental live TV show, watched by millions. It inspired a hit song, titled "Telstar," by a British band called the Tornados. The song opens with sounds of radio hiss, crackle, and electronic blips that give way to a synthesizer tune with an optimistic melody line. Even now it sounds futuristic. The song was an instant hit — the first British record to reach number one in the U.S., it ultimately sold five million copies worldwide. Ironically, Telstar was felled by atmospheric nuclear tests — the increases in radiation in the upper atmosphere overwhelmed its fragile transistors. According to the U.S. Space Objects Registry, its corpse is still in orbit.[18]

Telstar transformed TV's grasp of the world. In parallel with NASA's plans for the first lunar landing in 1969, a global network of communications satellites was placed into geosynchronous orbit. The network was ready just in time for the world to watch on TV as the Apollo 11 astronauts stepped onto the moon.[19]

With Telstar's deployment, the giant radio antenna for receiving signals from Echo 1 was no longer required. Penzias and Wilson, who both earned Ph.D.s in astronomy before joining Bell Labs, were allowed to use the antenna as a radio telescope, and they jumped at the chance. But as they collected their first data, they discovered a persistent hiss from the antenna, at microwave wavelengths, no matter which direction it faced on the sky. Try as they might, they could not get rid of the noise. Famously, they even tried scrubbing pigeon guano off the antenna, but the noise still came through.

Eventually, Penzias and Wilson's attention was drawn to a lecture that had just been given by James Peebles, a young theorist working with Robert Dicke nearby at Princeton, predicting radiation from the cosmos at just the wavelengths that would account for the hiss in their antenna. Unaware of Alpher, Gamow, and Herman's earlier calculations, Dicke and Peebles had reproduced their prediction of cosmic background radiation at a temperature of a few degrees Kelvin, corresponding to millimetre wavelengths — microwaves. So Penzias and Wilson phoned Dicke, then planning an experiment to search for the background radiation. As soon as Dicke put the phone down, he told his younger colleagues, "We've been scooped." Penzias and Wilson's discovery was the "shot heard around the world," immediately convincing almost all physicists that the universe had started with a hot big bang.

IN 1989, A DEDICATED NASA satellite called the Cosmic Background Explorer (COBE) took these background radiation measurements to a whole new level of precision. At the time, I was a professor at Princeton and had been working for a decade on how to test unified theories through cosmological observations. It was an exciting field, using the universe itself as the ultimate laboratory to test ideas about physics at extremely high energies, well beyond the reach of any conceivable experiment. However, I was worried about two things: first, that

many of the theories we were discussing were too contrived; and second, that the data was too limited to tell
which theory was right.

Then I went to the most dramatic seminar I have ever
attended. Held in the basement of the physics department in Princeton, it was the first unveiling of data
from the COBE satellite. The speaker was my colleague
in the department, David Wilkinson. Dave had been
part of Bob Dicke's original team that had been narrowly
scooped by Penzias and Wilson. By now he had become
one of the pioneers of experimental cosmology, having
made increasingly refined measurements of the background radiation, culminating in his involvement with
COBE. COBE marked a transition in the subject: before
it, almost nothing in cosmology was known to a better
accuracy than a factor of two. But after COBE, one precision measurement followed another. Theories are now
routinely proved wrong on the basis of discrepancies
with the data of only a few percent.

One of COBE's main experiments was the Far Infrared
Absolute Spectrophotometer (FIRAS), designed to measure the spectrum of the background radiation. The hot
big bang theory predicted that this spectrum should be
the Planck spectrum predicted by Max Planck when he
first proposed the quantization of light to describe radiation from a hot body. The background radiation was
emitted from the hot plasma of the early universe four
hundred thousand years after the expansion began, when
the temperature had fallen to a few thousand degrees. At

that time, the radiation took the form of red light. Once released, the radiation was stretched out a thousand-fold as it travelled across space, so that today most of the energy is carried in wavelengths of a few millimetres. Today, its spectrum is that of a hot body at just a few degrees Kelvin. In fact, to calibrate its measurements, FIRAS compared the sky with a perfectly black internal cavity whose temperature was dialled up and down to match that of the sky.

As Wilkinson put up a slide showing the measured spectrum, he said: "Here's a plot to bring tears to your eyes." There was an audible gasp from the audience. The data looked too good to be true. After just ten minutes of data, the measurements from the sky fit the Planck spectrum perfectly. By the time FIRAS completed its work, at every single frequency of radiation measured, the intensity matched the Planck spectrum to better than one part in a hundred thousand, and the temperature of the sky was measured at 2.725 degrees Kelvin. The measurements showed, in the most convincing way possible, that we live in a universe full of radiation left over from a hot big bang.

All I could think was, "Wow. The whole universe is shining down on us, saying, 'Quantum mechanics, quantum mechanics, quantum mechanics.'"

COBE HAD MORE IN store. Its measurements would indicate that quantum mechanics not only governed the cosmic radiation, it controlled the structure of the universe.

Following Penzias and Wilson's discovery, astro-physicists such as Peebles had been trying to work out how galaxies and other structures could have been formed from small density variations present in the very early universe. Einstein's equations predict how such variations grow with time. Places where there is greater density expand more slowly, and their gravity pulls more mass towards them. And places where there is less density expand more rapidly and empty out. In this way, gravity is the great sculptor of the universe, shaping planets, stars, galaxies, galaxy clusters, and all the other structures within the cosmos. Using Einstein's equations, we can trace this shaping process in reverse to figure out how the universe looked at earlier times. In other words, we can use the equations to retrace cosmic evolution. And then we can work out how the density variations should appear in the background radiation's temperature as it is mapped across the sky.

The COBE satellite carried another experiment called the Differential Microwave Radiometer (DMR). It was designed to scan for tiny differences in the temperature of the background radiation across the sky. The princi-pal investigator of that experiment was George Smoot of the University of California, Berkeley. Smoot, like Wilkinson, had pioneered measurements of the cosmic background radiation throughout his career.

I first met Smoot in 1991 at a summer school in Italy. Everyone wanted to know whether COBE's DMR experi-ment had detected any variations in the temperature across the sky. In Smoot's lectures, he showed a perfectly uniform map of the temperature. In private, he showed me slides of the actual measurements, which showed faint milky patches. Smoot was at that time still skeptical about whether the patches were real, and thought they might be an artefact of the experiment. And he had little confidence in theory: ever since the 1960s, theorists had been steadily lowering their predictions for the level of the expected temperature variations — from one part in a thousand, to a part in ten thousand, to one part in a hundred thousand, making them harder and harder for experimentalists like Smoot to detect.

The theorists had a reason to lower their predictions: they did so because of the growing evidence for dark matter. Dark matter's existence was first inferred in the 1930s by Swiss astronomer Fritz Zwicky, from observations of galaxies orbiting other galaxies in clusters. Their high velocities implied that more mass was present than could be accounted for by the visible stars. In the 1970s, U.S. astronomer Vera Rubin observed a similar effect in the outskirts of galaxies, where stars are seen to orbit so fast that if it were not for some extra gravity holding them in, they would escape from the galaxies.

Dark matter may consist of an unknown type of particle that does not interact with light or with ordinary matter. The only way to see these particles would be through their gravity. In the 1980s, astronomers found more evidence for dark matter by observing the bending of light by gravitational fields. This effect is called

Glenlair, James Clerk Maxwell's home in Dumfries and Galloway, Scotland. The house and grounds were Maxwell's childhood playground, providing many stimuli to the budding young scientist. See page 34.

A *field*, the key mathematical concept introduced in Maxwell's theory of electromagnetism. The arrows show the direction and strength of the field, and the grid of grey lines the coordinates, for each point in space. See page 41.

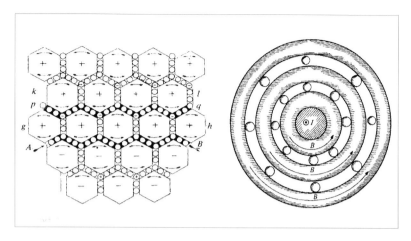

Maxwell's diagrams showing the machinery for magnetic fields and electric currents. On the left, the hexagonal cells are "vortices," representing a magnetic field. The particles between them carry an electric current. On the right is the magnetic field of a current in a wire. See pages 43–4.

The School of Athens by Raphael. Plato (left) and Aristotle (right) stand in the central arch. Seated at front centre, thinking and writing, is Heraclitus. To the left of him are Parmenides, Hypatia, Pythagoras, and Anaximander. At front right, using a compass, is Euclid. See pages 51–2.

The 1927 Fifth International Solvay Conference, held at the height of the quantum revolution which overturned the classical world-view. See pages 56–60.

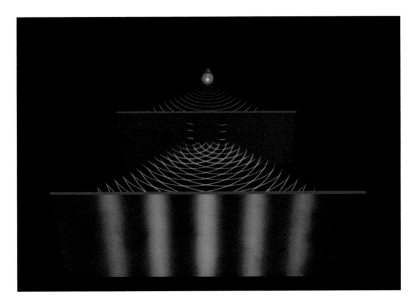

The double-slit experiment. A laser (top) shines light of a single wavelength on two slits in a partition (middle). The light waves from each slit spread out and interfere, producing a pattern of stripes on the screen (bottom). See pages 78–9.

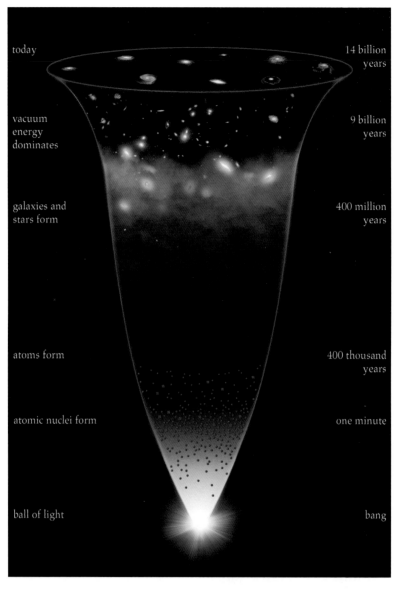

today · 14 billion years

vacuum energy dominates · 9 billion years

galaxies and stars form · 400 million years

atoms form · 400 thousand years

atomic nuclei form · one minute

ball of light · bang

Emergence of a region of the universe from the big bang. See pages 103–6.

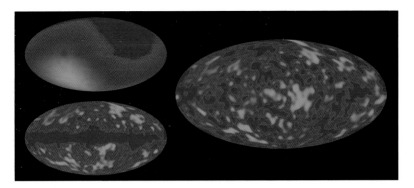

The temperature of the cosmic microwave background radiation, measured across the sky by the Differential Microwave Radiometer on NASA's COBE satellite. Hotter is red, colder is blue. At top left is the original picture, showing the asymmetry due to Earth's motion. Removing the effect of the motion produces the picture at lower left, showing the Milky Way as a band across the middle. Removing the galaxy's emission produces the picture at right, showing the primordial density variations in the universe. See pages 133–6.

The cluster of galaxies known as Abell 2218, two billion light years away from us. The stretched-out "arcs" of light are the images of galaxies behind the cluster, lensed and distorted by the cluster's gravitational field. The distortion can be used to measure the distribution of mass within the cluster, revealing the existence of a substantial quantity of "dark matter" in addition to the visible galaxies, stars, and hot gas. See pages 134–5.

The African Institute for Mathematical Sciences in Cape Town, South Africa.

A group of AIMS students at the 2008 launch of the Next Einstein Initiative, a plan to create fifteen AIMS centres across Africa within a decade. The two men in suits are Michael Griffin (left), head of NASA, and Mosibudi Mangena (right), the South African Minister for Science and Technology at the time. The woman in the head scarf at the picture's centre is Esra, from Sudan. The man at front right is Yves, from Cameroon. See pages 160–5.

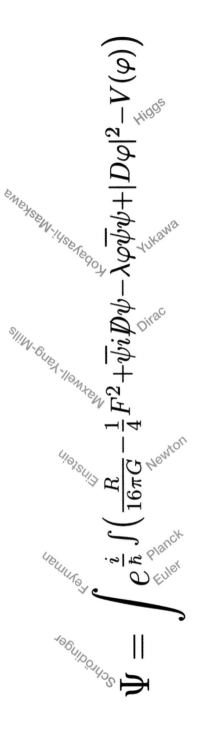

$$\Psi = \int e^{\frac{i}{\hbar} \int \left(\frac{R}{16\pi G} - \frac{1}{4}F^2 + \overline{\psi}i\slashed{D}\psi - \lambda\varphi\overline{\psi}\psi + |D\varphi|^2 - V(\varphi) \right)}$$

Feynman
Schrödinger
Euler
Planck
Einstein
Newton
Maxwell-Yang-Mills
Dirac
Kobayashi-Maskawa
Yukawa
Higgs

The formula that summarizes all the known laws of physics. See page 167–76.

The ATLAS detector at the Large Hadron Collider at CERN in Geneva, Switzerland. This giant apparatus (notice the person standing on the platform) is used to detect and analyze the spray of particles produced by the collision of two particle beams at its centre. See page 175.

Artist's impression of a typical collision at the Large Hadron Collider, revealing the existence of the Higgs boson. The discovery confirmed theorists' predictions, made a half-century ago, that the vacuum is permeated with a Higgs field. See pages 174–5.

"gravitational lensing" and is similar to the effect water has on light passing through it. Although water is perfectly transparent, you can tell that it is there because it bends the light and distorts the image of whatever is behind it. If you're in the bathtub and hold your hand up and look at the water droplets on the ends of your fingers, you'll see a distorted image of the room behind every drop. Astronomers have found similarly distorted images of distant galaxies behind galaxy clusters. And by figuring out how the light was bent, they can reconstruct the distribution of the dark matter within the clusters.

In the evolution of the universe, dark matter would have played a very important role, assisting the formation of galaxies by providing an extra gravitational pull. In many ways, dark matter forms a kind of cosmic backbone, holding the other matter together in the universe. Dark matter's gravitational pull lowered the minimal level of density variations required in the early plasma to form galaxies by today, bringing it down to one part in a hundred thousand. It was very hard to see how galaxies could have formed from density variations any smaller than that. So as COBE reached this crucial sensitivity level, it was also reaching the smallest level of variations that could possibly have formed our current universe's structure.

Fortunately, the milky patches Smoot showed me turned out to be real. The temperature across the sky really does vary by a part in a hundred thousand, and in an incredibly simple way. The observations matched precisely expectations from inflationary theories like those discussed in the workshop I'd attended in Cambridge in 1982. When DMR's result was announced a decade later, Stephen Hawking was quoted as saying the finding was the greatest discovery of the twentieth century, and perhaps of all time. Although his comment was hyperbolic, there was good reason to be excited.

AS MEASUREMENTS OF THE universe became far more accurate and extended to a greater and greater volume, it became possible to envisage using the entire visible universe as a giant laboratory. The big bang is the ultimate high-energy experiment. We and everything around us are its consequence. So understanding the very early universe allows us to probe physics at the shortest distances and the very highest energies. Equally, looking at today's universe allows us to probe the largest distances and the very lowest energies. And in this probing, cosmology made another of the greatest discoveries of twentieth-century physics, the full implications of which we are still struggling to understand.

I have already mentioned Einstein's earliest cosmological model and how it included the cosmological term. The model was a failure, but the idea of the cosmological term was a good one. In fact, it was Lemaître who persisted with it, arguing that it was a plausible addition to Einstein's theory and should be thought of as a special, simple kind of matter that could be expected to be present in the universe.

Later on, it was realized that the cosmological term represents the energy per unit volume of empty space, what we now call the "vacuum energy." It is the very simplest form of energy, being completely uniform across space and appearing exactly the same to any observer.[20]

For any physical process that does not involve gravity, the vacuum energy makes no difference. It is just there all the time as an unchanging backdrop. The only way to detect it is through its gravity, and the best way to do that is to look at as big a chunk of it as possible. Of course, the biggest piece of space we have is the entire visible universe. By watching the expansion history of this whole region, you can directly measure the vacuum energy's gravity.

In 1998, the High-Z Supernova Search Team and the Supernova Cosmology Project — two international ventures led by Australian National University's Brian Schmidt, Johns Hopkins's Adam Reiss, and Saul Perlmutter at the University of California, Berkeley — measured the brightness and recession speeds of exploding stars called "supernovae," which are so bright that they are visible even in very distant galaxies. Their measurements showed conclusively that the expansion of the universe has begun to speed up, pointing to a positive vacuum energy. Perlmutter, Reiss, and Schmidt shared the 2011 Nobel Prize for this discovery. An article on the Nobel Prize website describes the vacuum energy's repulsive effect: "It was as if you threw a ball in the air and it kept speeding away until it was out of sight."

The discovery of a positive vacuum energy was a key step towards settling the makeup of the universe. In today's universe, the vacuum energy accounts for 73 percent of the total energy. Dark matter accounts for 22 percent, ordinary matter like atoms and molecules 5 percent, and radiation a tiny fraction of a percent. Dark matter, ordinary matter, and radiation emerged from the early universe with gentle ripples, at a level of one part in a hundred thousand on all scales, which seeded the creation of galaxies. This "concordance model" of the universe has, over the past decade, enjoyed one success after another as all kinds of observations fell in line with it. So far, there are no clouds on its horizon.

So far, we have no idea how to use dark matter or vacuum energy, but it is tempting to speculate that, some day, they might provide a ready source of fuel that we could use to travel across space. In fact, special relativity makes the universe far easier to explore than you might think at first sight. Lorentz contraction means that, for space travellers, the universe can be crossed in a relatively short period of time.

Think about a spaceship that escapes from Earth's gravity and continues thereafter at one g—that is, with the same acceleration that a falling object has on Earth. This would be quite comfortable for the space travellers, since they would feel an artificial gravity of just the same strength as gravity on Earth. After a year or so, the spaceship would approach the speed of light, and it would get closer and closer to light speed as time went on. As the universe flashed by, Lorentz contraction would make it appear more and more compressed in the direction of travel. After just twenty-three years of the space travellers' time, they would have crossed the whole region of space we can currently see. Of course, it would take another twenty-three years to slow down the ship so they could get off and explore. And due to time dilation, billions of years would have elapsed back on Earth.

For now, these prospects seem very distant. But if history is anything to go on, they may be nearer than we think.

· · ·

THE PICTURE OF COSMOLOGY I have described has been remarkably successful. It has accommodated every new observation, and it now forms the foundation for more detailed studies of the formation of galaxies, stars, and planets in the universe. We have built the science of the universe. Why, then, aren't we theorists satisfied?

The problem, as I mentioned earlier, is that working models of inflation are no more beautiful today than they were in 1982. Inflationary models do not explain what happened just before inflation, or how the universe emerged from a cosmic singularity. It is simply imagined that the universe somehow sprang into being filled with inflationary energy.

And the more we have learned about unification, the more contrived inflationary models appear. In addition to assuming the universe started out inflating, the models' parameters have to be adjusted to extremely small values in order to fit the data. There is no shortage of inflationary models: there are thousands of them. The problem is that they are all *ad hoc*, and it would be impossible to distinguish many of them from each other through observations.

The small patch of space that started out inflating was assumed to be filled with inflationary energy at extremely high density. The inflationary energy will ultimately decay, at the end of inflation, into matter and radiation. However, in any realistic model there must also, at the end, be a tiny residue of vacuum energy to explain what we see today. In any inflationary model, one can ask: what is the ratio of the assumed inflationary energy density at the beginning of inflation to the vacuum energy density we now measure in the universe? That ratio must appear in the description of the model: it is a vast number, typically a googol (a one with a hundred zeros after it) or so. In every known model so far, this number is just assumed, or picked from a vastly greater number of models according to a principle we do not yet understand.

Every inflationary model suffers from this fine-tuning difficulty in addition to all of the other problems. Remember, inflationary theory was invented in order to explain the peculiar fine-tuned, smooth, and flat initial conditions required in the ball of light at the start of the hot big bang. But now we find that inflation is itself based on a strange and artificial initial condition in which the inflationary energy takes a vast value for no apparent reason.

You can think of inflationary energy as a highly compressed spring, like the one you compress when you start a game of pinball. To get the ball going as quickly as possible, you must condense the spring as much as you can. This is similar to what you need for inflation: you need an enormous density of inflationary energy, compressed into a tiny region of space. But how likely would it be for you to come across a pinball machine that spontaneously shot the ball up into the machine? It is possible that the random vibrations of the spring and the collisions of all the surrounding air molecules conspire to kick the ball up, but it is extremely unlikely. The conditions required to initiate inflation are, it turns out, vastly more improbable.

It is true that the total amount of energy required to get inflation going isn't so great, and Guth relies on this in his "free lunch" argument. But energy is not a good measure of how extreme such an inflating region is, because energy is not conserved when space is expanding, and certainly not during inflation. There is a known measure of the rarity of inflationary initial conditions, known as the "gravitational entropy." Roughly speaking, it allows you to ask how unusual it would be to find a patch of the universe with inflating initial conditions. The answer is that inflating initial conditions would be expected around one time in 10 raised to the power of 10 raised to the power of 120. That is an extremely small probability, and points to a serious problem with the inflationary hypothesis.

The most serious attempt so far to describe the initial conditions required for inflation was made by James Hartle and Stephen Hawking, building on earlier ideas of the Russian cosmologist Alexander Vilenkin. They noticed that because inflationary energy is repulsive, it is possible for universes to avoid a singularity. They considered a curved universe where space takes the form of a small three-dimensional sphere, and showed that such a universe, if you filled it with inflationary energy, could be started out set at some time in a static condition. If you followed time forward, the universe would grow exponentially in size. Likewise, if you followed time backward it would also grow exponentially. So if you followed time forward from some point long before the universe was static, you would see the universe first shrink and then "bounce" from contraction to expansion.

The effect is like a bouncing ball. Imagine someone shows you a time-lapse movie whose first frame shows the ball at the moment of the bounce, squished up against the floor. As time proceeds, the ball pushes itself off from the floor and becomes spherical again. Now they play the movie again, but this time running backward in time from the moment of the bounce. There is no difference! The laws of physics are unchanged if time is reversed: going backward in time, the ball will do just as it did going forward. Of course, if they show the whole movie, starting at some time well before the bounce, you will see the ball start out spherical as it hits the floor, squish up against the ground, and then unsquish itself as it bounces off.

Hartle and Hawking, following Vilenkin, employ the powerful mathematical trick of imaginary time, which I explained at the end of the previous chapter. Hawking had applied this trick successfully to black holes, showing that they possess a tiny temperature and emit radiation called "Hawking radiation." Now he and Hartle tried to apply it to the beginning of the universe. If you follow time back to the bounce, you can use the imaginary number i to change time into another direction of space. And now, it turns out, with four space dimensions the geometry of the universe can be "rounded off" smoothly, with no singularity. Hartle and Hawking called this idea the "no boundary" proposal because in their picture, the universe near its beginning would be a four-dimensional closed surface, like the surface of a sphere, with no boundary. In spirit, their idea is highly reminiscent of Lemaître's "Primeval Atom" proposal.

When I moved to Cambridge, in 1996, I worked with Hawking and a number of our Ph.D. students to develop the predictions of the Hartle–Hawking proposal for general theories of inflation. We showed that in such theories, the universe in the imaginary time region could generally be described as a deformed four-dimensional sphere, a configuration that became known as the "Hawking–Turok instanton." It turns out you can work out all the observational quantities within this region, and then follow them forward to the moment of the "bounce" and then into the normal, expanding region of spacetime, where they determine what observers would actually see.

A beautiful feature of the Hartle–Hawking proposal is that it does not impose an arbitrary initial condition on the laws of physics. Instead, the laws themselves define their own quantum starting point. According to the Hartle–Hawking proposal, the universe can start out with *any* value for the inflationary energy. Their proposal predicts the probability for each one of these possible starting values. It turns out that this calculation agrees with the estimate of gravitational entropy I mentioned earlier: the probability of getting realistic inflationary initial conditions is around one in 10 to the power of 10 to the power of 120. The most probable starting point, by far, is the one with the smallest possible value of the inflationary energy, that is, today's vacuum energy. There would be no period of inflation, no matter or radiation. Hartle and Hawking's proposal is a wonderful theory, but at least in the most straightforward interpretation, it predicts an empty universe.

Hartle and Hawking and their collaborator Thomas Hertog, of the University of Leuven, propose to avoid this prediction by invoking the "anthropic principle" — the idea that one should select universes according to their ability to form galaxies and life.

It is not a new notion that the properties of the universe around us were somehow "selected" by the fact that we are here. The idea has grown increasingly popular as theory has found it more and more difficult to explain the specific observed properties of the universe. The problem is that the anthropic arguments are vague: in order to make them meaningful, one needs a theory of the set of possible universes and also the precise condition for us to be located in one of them. Neither of these requirements are yet close to being met. Nevertheless, Hartle, Hawking, and Hertog argue that even if *a priori* an empty universe is the most likely, the predictions of the Hartle–Hawking proposal, supplemented by anthropic selection, are consistent with what we observe. In principle, I have no objection to this kind of argument, as long as it can really be carried through.

However, a realistic universe like ours has a minuscule *a priori* probability in this setup, of one in 10 raised to the power of 10 raised to the power of 120 (the same tiny number mentioned earlier). Anthropic selection has to eliminate *all* of the other possible universes, and this seems an extremely tall order. A universe in which ours was the only galaxy, surrounded by empty space, would seem to be quite capable of supporting us. And, according to the Hartle–Hawking proposal, it would be vastly more likely than the universe we observe, which is teeming with galaxies (Hartle, Hawking, and Hertog exclude such a universe by *fiat* in their discussion). When the *a priori* probabilities are so heavily stacked against a universe like ours, as they are with the Hartle–Hawking proposal, it seems to me very unlikely that anthropic arguments will save the day.

A THEORY THAT PREDICTS a universe like ours *a priori*, without any need for anthropic selection, would seem vastly preferred. Even if anthropic selection *could* rescue Hartle and Hawking's theory (which seems to me unlikely), the non-anthropic theory would be statistically favoured over the anthropic one by a huge factor, of 10 raised to the power of 10 raised to the power of 120.

For the past decade, with Paul Steinhardt of Princeton University and other collaborators, I have been trying to develop such theories as an alternative to inflation. Our starting point is to tackle the big bang singularity. What if it was not the beginning of time, but instead was a gateway to a pre–big bang universe? If there was a universe like ours before the singularity, could it have directly produced the initial ball of light, and if it did, would there be any need for a period of inflation?

Most of the universe today is very smooth and uniform on scales of a millimetre, and we have no problem understanding why. Matter and radiation tend to spread themselves out through space, and the vacuum energy is completely uniform anyway. Let us imagine following our universe forward into the future. The galaxies and all the radiation will be diluted away by the expansion: the universe will become a cold, empty place, dominated by the vacuum energy. Now imagine that for some reason the vacuum energy is not absolutely stable. It could start to slowly decay, tens of billions of years into our future. We can easily build mathematical models where it declines in this way, becoming smaller and smaller and then going negative. Its repulsive gravity would become attractive, and the universe would start to collapse.

When we studied this idea, we discovered that the pressure of the unstable energy would become large and positive, and it would quickly dominate everything else. As the universe collapsed, this large positive pressure would quickly make the universe very smooth and flat. When the universe shrunk down to zero size, it would hit a singularity. Then, plausibly, the universe would rebound, fill with radiation, and start expanding again. In fact, immediately after the bounce we would have conditions just like those in our millimetre-sized ball of light: the very initial conditions that were needed to explain the hot big bang.

Much to our surprise, we found that during the collapse initiated by the unstable vacuum energy, our high-pressure matter develops quantum variations of exactly the form required to fit observations. So in this picture, we can reproduce inflationary theory's successes, but with no need for initial inflationary conditions.

Our scenario is far more ambitious than inflation in attempting to incorporate and explain the big bang singularity. We have based our attempts on M-theory, a promising but still developing framework for unifying all the laws of physics. M-theory is the most mathematical theory in all of physics, and I won't even try to describe it here.

Einstein used the mathematics of curved space to describe the universe. M-theory uses the same mathematics to describe everything *within* the universe as well.

For example, string theory, which is a part of M-theory, describes a set of one-dimensional universes — pieces of string — moving within higher-dimensional space. Some strings describe force-carrier particles like photons, gluons, or gravitons, while others describe matter particles like electrons, quarks, or neutrinos. As well as strings, M-theory includes two-dimensional universes, called "membranes," and three-dimensional universes, called "3-branes," and so on. According to M-theory, all of these smaller universes are embedded within a universe with ten space dimensions and one time dimension, which seems more than rich enough to contain everything we see.

In the best current versions of M-theory, three of the ten space dimensions — the familiar dimensions of space — are very large, while the remaining seven are very small. Six of them are curled up in a tiny little ball whose size and shape determine the pattern of particles and forces we see at low energies. And the seventh, most mysterious dimension, known as the "M-theory dimension," is just a tiny gap between two three-dimensional parallel worlds.

Until our work, most M-theorists interested in explaining the laws of particle physics today had assumed that all the extra, hidden dimensions of space were static. Our new insight was to realize that the extra dimensions could change near the big bang, and that the higher-dimensional setting would cast the big bang singularity in a new light.

What we found was that, according to M-theory, the big bang was just a collision between the two three-dimensional worlds living at the end of the M-theory dimension. And when these worlds collide, they do *not* shrink to a point—from the point of view of M-theory, the three-dimensional worlds are like two giant parallel plates running into each other. What our work showed was that, within M-theory, the big bang singularity was, after all, not as singular as it might first appear, and most physical quantities, like the density of matter and radiation, remain completely finite.

Recently, we have discovered another, very powerful way to describe how the universe passes through the singularity, which turns out not to rely on all the details of M-theory. The trick uses the same idea of imaginary time which Hartle and Hawking used to describe the beginning of spacetime. But now we use imaginary time to circumvent the singularity, passing from a pre-bang collapsing universe to a post-bang expanding universe like the one we see today. We are close to finding a consistent and unique description of this process and to opening a new window on the pre-bang world.

If the universe can pass through a singularity once, then it can do so again and again. We have developed the picture into a cyclic universe scenario, consisting of an infinite sequence of big bangs, each followed by expansion and then collapse, with the universe growing in size and producing more and more matter and radiation in every cycle. In this picture of the universe, space is infinite and so too is time: there is no beginning and there is no end. We called this an "endless universe."[21] A cyclical universe model may, in its evolution, settle down to a state in which it repeats the same evolution, in its broad properties, over and over again. In this way, the vast majority of space would possess the physical properties of the universe we see. There would be no need for anthropic arguments, and the theoretical predictions would be clearer.

IF THERE IS ONE rule in basic physics, I would say it is "in the long run, crime does not pay." Cosmology in the twentieth century was, by and large, based on ignoring the big bang singularity. Yet the singularity represents a serious flaw in the theory, one which it is possible to ignore only by making arbitrary assumptions, which, in the end, may have little foundation. By continuing to ignore the singularity, we are in danger of building castles of sand. The singularity may just be our greatest clue as to where the universe really came from. Our work on the cyclic universe model has shown that all of the successes of the inflationary model can be reproduced in a universe that passes through the singularity without undergoing any inflation at all.

The competition between the cyclic and inflationary universe models highlights one of the most basic questions in cosmology: did the universe begin? There are only two possible answers: yes or no. The inflationary and cyclic scenarios provide examples of each possibility.

The two theories could not be more different: inflation assumes a huge burst of exponential expansion, whereas the cyclic model assumes a long period of slow collapse. Both models have their weak points, mathematically, and time will tell whether these are resolved or prove fatal. Most exciting, the models make different observational predictions which can be tested in the not-too-distant future.

At the time of writing, the European Space Agency's Planck satellite is deep in space, mapping the cosmic background radiation with unprecedented precision. I have already discussed how inflation can create density variations in the universe. The same mechanism — the burst of inflationary expansion — amplifies tiny quantum gravitational waves into giant, long-wavelength ripples in spacetime, which could be detectable today. One of the Planck satellite's main goals is to detect these very long wavelength gravitational waves though their effects on the temperature and polarization of the cosmic background radiation across the sky. In many inflationary models, including the simplest ones, the effect is large enough to be observed.

Throughout his career, Stephen Hawking has enjoyed making bets. It's a great way of focusing attention on a problem and encouraging people to think about it. When I gave my first talk on the cyclic model in Cambridge, I emphasized that it could be observationally distinguished from inflation because, unlike inflation, it did *not* produce long wavelength gravitational waves. Stephen immediately bet me that the Planck satellite would see the signal of inflationary gravitational waves. I accepted at once, and offered to make the bet at even odds for any sum he would care to name. So far we haven't agreed on the terms, but we will do so before Planck announces its result, which may be as soon as 2013. Another leading inflationary theorist, Eva Silverstein of Stanford University, has agreed to a similar, though more cautious bet: the winner will get either a pair of ice skates (from me, in Canada) or a pair of rollerblades (she being from California).

. . .

LOOKING BACK OVER PAST millennia, we have to feel privileged to be alive at a time when such profound questions about the universe are being tackled, and when the answers seem finally within reach. In ancient Greece, there was a debate that in many ways prefigured the current inflationary/cyclic competition. Parmenides of Elea held the view — later echoed by Plato — that ideas are real and sensations are illusory, precisely the opposite of the views later espoused by David Hume. If thought is reality, then anything one can conceive of must exist. Parmenides reasoned that since you cannot think of something not existing without first thinking of the thing itself, then it is logically impossible for anything to come into existence. Hence he believed all change must be an illusion: everything that happens must already be implicit in the world. This is a fairly accurate description of Hartle and Hawking's "no boundary" proposal. To work out the predictions of their proposal, one works in "imaginary time" — in the primordial, quantum region of spacetime where everything that happens subsequently in the universe is implicit, and one continues the predictions into real time to see what they mean for today's observations.

On the other hand, Heraclitus of Ephesus, like Anaximander before him, held the opposite point of view. "All is flux" was his dictum: the world is in constant tension between its opposing tendencies. Everything changes and nothing endures. The goal of philosophy, he argued, is to understand how things change, both in society and in the universe. Starting with Zeno, the Stoic philosophers introduced the concept of *ekpyrosis*, meaning "out of fire," to describe how the universe begins and ends in a giant conflagration, with a period of normal evolution in between. In his treatise *On the Nature of the Gods*, Cicero explains, "There will ultimately occur a conflagration of the whole world...nothing will remain but fire, by which, as a living being and a god, once again a new world may be created and the ordered universe restored as before."[22] There were similar ideas in ancient Hindu cosmology, which presented a detailed cyclic history of the universe.

In the Middle Ages, the idea of a cyclic universe became less popular as Christianity took hold and the biblical explanation of a "beginning" became the norm.

Nevertheless, cyclic ideas regularly appeared — Edgar Allan Poe wrote an essay titled "Eureka" that proposed a universe resembling the ancient ekpyrotic picture. And the German philosopher Friedrich Nietzsche also advocated a repeating universe. He argued that since there can be no end to time and there are only a finite number of events that can occur, then everything now existing must recur, again and again for eternity. Nietzsche's model of "eternal recurrence" was popular in the late nineteenth century.

In fact, Georges Lemaître, even as he worked on the idea of a "quantum beginning," commented favourably on Friedmann's oscillating cosmological solutions. In 1933, he wrote that these cyclic models possessed "an indisputable poetic charm and make one think of the phoenix of the legend."[23]

For now, we stand on the verge of major progress in cosmology. Both theory and observation are tackling the big bang in our past, and they will determine whether it was really the beginning of everything or merely the latest in a series of bangs, each one of which produced a universe like ours. They are also tackling the deep puzzle of the vacuum energy that now dominates the universe, and which will be overwhelmingly dominant in the future. What is it composed of, and can we access it? Will it last forever? Will the exponential expansion it drives dilute away all of the stars and galaxies that surround us and lead to a vacuous, cold eternity? Or will the vacuum energy itself seed the next bang? These questions have entered the realm of science and of scientific observation. I, for one, cannot wait for the answers.

FOUR

THE WORLD IN AN EQUATION

"If you are receptive and humble, mathematics will
lead you by the hand. Again and again, when I have
been at a loss how to proceed, I have just had to wait
until I have felt the mathematics lead me by the hand.
It has led me along an unexpected path, a path where
new vistas open up, a path leading to new territory,
where one can set up a base of operations, from which
one can survey the surroundings and plan future
progress." — Paul Dirac, 1975[1]

USING THE MOST POWERFUL radio telescope on earth,
astronomers have just detected an encrypted signal
emanating from Vega, one of the brightest stars in the
sky and about twenty-five light years away. The message
contains instructions for building a machine to teleport
five human beings across space to meet with the extra-
terrestrials. After an intense international search, world

leaders select the five delegates. Among them is the brilliant young Nigerian physicist Dr. Abonnema Eda, who
has just won the Nobel Prize for discovering the theory
of superunification, combining all known physics into a
single, unified picture.

The storyline comes from Carl Sagan's 1985 novel
Contact, later made into a movie starring Jodie Foster.
Sagan was a renowned U.S. astronomer and, with his
TV series *Cosmos*, one of science's greatest popularizers.
In casting Eda as a hero in his novel, Sagan was making two points: first, that discovering the basic laws of
the universe is a global, cross-cultural field of research.
People from every country are fascinated by the same
questions about how the world works. Second, genius
knows no national boundaries. Although Africa has so
far been woefully under-represented in the history of
physics, like other disadvantaged regions, in the future
it may be a source of incredible talent. Science benefits
greatly from a diversity of cultures, each bringing a new
stimulus of energy and ideas.

OVER THE PAST DECADE I have led a dual existence. On
the one hand, I have been trying to understand how to
describe the beginning and the far future of the universe.
On the other, I have been fascinated by the problem of
how to enable young people to enter science, especially
in the developing world.

My interest is rooted in my African origins. As I
described earlier, I was born in South Africa, where

my parents were imprisoned for resisting the apartheid regime. Upon their release, we left as refugees, first to East Africa and then the U.K. When I was seventeen, I returned to Africa to teach for a year in a village mission school in Lesotho, a tiny, landlocked country surrounded by South Africa. Lesotho is one of the poorest nations on Earth: 80 percent of the jobs available are migrant labour, mostly in the mines over the border. In the village of Makhakhe, where I worked, I met many wonderful people and great kids with loads of potential but zero opportunity. No matter how bright or talented they were, they would never have the chances I'd had. A clerical position in the mines was the height of their aspiration.

The kids in the mission school were eager, responsive, and bright. But by and large, the education they'd received consisted of rote learning: memorizing times tables, copying notes from the blackboard and reproducing them in exams. They had no real experience of figuring things out or learning to think for themselves. School was a dry exercise you had to submit to: its sole purpose was to get a certificate. The teachers had been taught that way themselves, and they perpetuated a cycle of harsh discipline and low expectations.

I tried to take the kids outside as often as possible, to try to connect what we did in class with the real world. One day I asked them to estimate the height of the school building. I expected them to put a ruler next to the wall, stand back and size it up with finger and thumb, and make an estimate of the wall's height. But there was one boy, very small for his age and the son of one of the poorest families in the village, who was scribbling with chalk on the pavement. A bit annoyed, I said, "What are you doing? I want you to estimate the height of the building." He said, "I measured the height of a brick. Then I counted the number of bricks and now I'm multiplying." Well, needless to say, I hadn't thought of that!

People often surprised me with their enthusiasms and interests. Watching a soccer match at the school one day, I sat next to a miner, at home on his annual leave. He told me, "There's only one thing that I really loved at school: Shakespeare." And he recited some lines to me. Many similar experiences convinced me of the vast potential for intellectual development which is sorely needed for the continent's progress.

Evolution was not on the school curriculum, because the church objected, but we nevertheless had excellent classroom discussions about it. Most African children are unaware of modern scientific discoveries showing how *Homo sapiens* originated in Africa around two hundred thousand years ago and began to migrate out of Africa between seventy thousand and fifty thousand years ago. I believe they could draw motivation from learning how humankind, and mathematics, and music and art, arose in Africa. Instead, young Africans are all too often made to feel like helpless bystanders, with every advance happening elsewhere in the world.

With the end of apartheid in 1994, my parents were allowed to return to South Africa. Both won election as members of the new parliament for the African National Congress, alongside Nelson and Winnie Mandela. They kept saying to me, "Can't you come back and help in some way?" At the time, I was too busy with my own scientific career. Eventually, in 2001, I took a leave from my position at Cambridge to visit the University of Cape Town, near where my parents lived. Most of the time, I was pursuing new cosmological theories, like the cyclic universe scenario. But I took time out from my research to meet with colleagues and discuss what we might do to help speed Africa's scientific development.

In these conversations, it quickly emerged that Africa's deficiency in maths is a serious problem. There is an acute scarcity of engineers, computer scientists, and statisticians, making it impossible for industry to innovate or, more generally, for governments to make well-informed decisions regarding health, education, industry, transport, or natural resources. African countries are highly dependent on the outside world; they export raw and unprocessed commodities, and import manufactured goods and packaged food. Cellphones have transformed the lives of many Africans, but none are yet made in Africa. If Africa is to become self-sufficient, it urgently needs to develop its own community of skilled people and scientists to adapt and invent the technologies that will allow it to catch up to the rest of the world.

And so we came up with the idea of setting up a centre

called the African Institute for Mathematical Sciences, or AIMS, which would serve the continent. The idea was very simple: to recruit the brightest students from across Africa and the best lecturers from around the world for a program designed to turn Africa's top graduates into confident thinkers and problem solvers, skilled in a range of techniques like mathematical modelling, data analysis, and computing. We would provide exposure to many scientific fields of great current relevance to Africa, like epidemiology, resource and climate modelling, and communications, but we would also mix in foundational topics like basic physics and pure mathematics.

Above all, we wanted to create a centre with a culture of excellence and a commitment to Africa's development. The goal of the institute would be to open doors; encourage students to explore and develop their interests; discover which fields excited them the most; and assist them in finding opportunities. AIMS would help them forge their path to becoming scientists, technologists, educators, advisors, and innovators contributing to their continent's growth.

With my parents' encouragement, my brothers and I used a small family inheritance to purchase a disused, derelict hotel. It's an elegant eighty-room, 1920s art deco building in a seaside suburb of Cape Town. Then, with the support of my colleagues in Cambridge, we formed a partnership involving Cambridge and Oxford universities in the U.K., Orsay in France, and the three main universities in Cape Town. We recruited a South African nuclear physicist, Professor Fritz Hahne, as the institute's first director. I persuaded many of my academic colleagues to teach for three weeks each, and we advertised the program across Africa by emailing academic contacts and by sending posters to universities. In 2003, we opened AIMS, receiving twenty-eight students from ten different countries.

AIMS was launched as an experiment. We started it out of belief and commitment but had no real idea how or whether it would work. Most of us working to develop the project were academics, with little experience of creating an institution from scratch. For everyone involved, it was a wonderful learning experience. What we discovered was that when you take students from across a continent as diverse as Africa and put them together with some of the best teachers in the world, sparks fly.

South Africa has a strong, largely white, scientific community. Many of the local academics said to us, "Are you sure you really want to do this? You're going to be spending all your time on remedial teaching. These students won't know a thing." How quickly AIMS proved them wrong! What the students lack in preparation, they make up for with motivation. Many students have had to overcome incredible difficulties, whether poverty, war, or loss of family members. These experiences make them value life, and the opportunity AIMS provides them, even more. They work harder than any students I have ever seen, and they feel and behave as if the world is opening up in front of them.

Teaching at AIMS is an unforgettable experience. One feels a tremendous obligation to teach clearly and well, because the students really want to learn and need to learn fast. There is a profound shared sense at AIMS that we are participating in the scientific transformation of a continent. And we are filled with the belief that when young Africans are given the chance to contribute, they will astonish everyone.

Take Yves, from the heart of Cameroon, from a peasant family with nine children. His parents could afford to send only one of them to university. Yves was the lucky one, and he is determined to live up to the opportunity and to prove what he can do. After graduating from AIMS, he took a Ph.D. in pure mathematics. Soon after, he won the prize for the best Ph.D. student presentation at the annual South African Mathematical Society conference. What an achievement, and what a powerful symbol that someone who comes from a poor family in an African village can become a leading young scientist.

In just nine years of existence, AIMS has graduated nearly 450 students from thirty-one different African countries. Thirty percent of them are women. Most come from disadvantaged backgrounds, and almost all have gone on to successful careers in research, academia, enterprise, industry, and government. Their success sends a powerful message of hope which undermines prejudice and inspires countless others. What more cost-effective investment could one possibly make in the future of a continent?

Ever since we started AIMS in Cape Town, our dream was to create a network of centres, providing outstanding scientific education across the continent. Specifically, our plan was to open fifteen AIMS centres. Each should serve as a beacon, a jewel in the local scientific and educational firmament, helping to spark a transformation in young people's aspirations and opportunities.

In 2008, I was invited by the TED organization to make "a wish to change the world" at their annual conference in California, attended by some of the most influential people in Silicon Valley. My wish was "that you help us unlock and nurture scientific talent across Africa, so that within our lifetimes we are celebrating an African Einstein." Using Einstein's name in this way is not something a theoretical physicist does lightly, and I must admit that I was nervous. Before I did, I sounded out some of my most critical physicist colleagues to see how they would respond. To my delight, they were unreservedly enthusiastic. Science needs more Einsteins, and it needs Africa's participation.

The idea of using Einstein's name in our slogan came from from another remarkable AIMS student. Esra comes from Darfur in western Sudan. Her family suffered from the genocide there, in which tens of thousands were murdered and millions displaced. Esra was doing physics at Khartoum University before she made her way to AIMS. In spite of her family and community's desperate problems back home, she somehow manages to remain cheerful.

One evening at AIMS, I lectured on cosmology. As usual, there was a lot of very animated discussion. At one point, I showed the Einstein equation for the universe and, as an aside, said, "Of course, we hope that among you there will be another Einstein." The next day, a potential donor was coming to visit, and we asked a number of the students, including Esra, to speak. She ended her short, moving speech with the words: "We want the next Einstein to be African." So when TED called me, a few weeks later, and asked if I had a wish, I knew immediately what it would be.

The slogan is deliberately intended to reframe the goals of international aid and development. Instead of seeing Africa as a problem continent, beset by war, corruption, poverty, and disease, and deserving of our charity, let us see it for what it can and should be: one of the most beautiful places on Earth, filled with talented people. Africa is an enormous asset and opportunity for the world. For too long, Africa has been exploited for its diamonds, gold, and oil. But the future will be all about Africa's people. We need to believe in them and what they are capable of.

Modern society is built upon science and scientific ways of thinking. These are our most precious possessions and the most valuable things we can share. The training of African scientists, mathematicians, engineers, doctors, technologists, teachers, and other skilled people should be given the highest priority. And this should be done not in a patronizing way but in a spirit of mutual respect and mutual benefit. We need to see Africa for what it is: the world's greatest untapped pool of scientific talent.

In encouraging young Africans to aim for the heights of intellectual accomplishment, we will give them the courage and motivation to pursue advanced technical skills. Among them will be not only scientists, but also people entering government or creating new enterprises: the African Gateses, Brins, and Pages of the future.

Last year, we opened our second centre, AIMS-Sénégal, in a beautiful coastal nature reserve just south of Dakar. This year, the third AIMS centre opened in Ghana, in another attractive seaside location. AIMS-Ethiopia will be next.

AIMS now receives nearly five hundred applications per year, and our graduates are already having a big impact in many scientific fields, from biosciences to natural resources and materials science, engineering, information technologies, and mathematical finance, as well as many areas of pure maths and physics. They are blazing a trail for thousands more to follow. AIMS will, we hope, serve as the seed for building great science across Africa.

Very recently, South Africa won the international competition to host what will be the world's largest radio telescope — the Square Kilometre Array (SKA). The array will span 5,000 kilometres and include countries from Namibia to Kenya and Madagascar. It will be one of the most advanced scientific facilities in the world, placing Africa at the leading edge of science and helping to inspire a new generation of young African scientists. Among them may well be an Abonnema Eda.

. . .

OVER THE COURSE OF the twentieth century, in pursuit of superunification, physicists have produced a one-line formula summarizing all known physics: in other words, the world in an equation. Much of it is written in Greek, in homage to the ancients. The mathematics of the Pythagoreans, and most likely the ancient Sumerians and Egyptians before them, lies at its heart. Their beliefs in the power of mathematical reasoning and the fundamental simplicity of nature have been vindicated to an extent that would surely have delighted them.

This magic formula's accuracy and its reach, from the tiniest subatomic scales to the entire visible universe, is without equal in all of science. It was deciphered through the combined insights and labours of many people from all over the world. The formula tells us that the world operates according to simple, powerful principles that we can understand. And in this, it tells us who we are: creators of explanatory knowledge. It is this ability that has brought us to where we are and will determine our future.

Every atom or molecule or quantum of light, right across the universe, follows the magic formula. The incredible reliability of physical laws is what allows us to build computers, smartphones, the internet, and all the rest of modern technology. But the universe is not like a machine or a digital computer. It operates on quantum laws whose full meaning and implications we are still discovering. According to these laws, we are not irrelevant bystanders. On the contrary, what we see depends upon what we decide to observe. Unlike classical physics, quantum physics allows for, but does not yet explain, an element of free will.

Let us start from the left of the formula with Schrödinger's wavefunction, Ψ, the capital Greek letter pronounced *psi*. Every possible state of the world is represented by a number, which you get using Ψ. But it isn't an ordinary number; it involves the mysterious number i, the square root of minus one, which we encountered in Chapter Two. Numbers like this are called "complex numbers." They are unfamiliar, because we don't use them for counting or measuring. But they are very useful in mathematics, and they are central to the inner workings of quantum theory.

A complex number has an ordinary number part and an imaginary number part telling you how much of i it contains. The Pythagorean theorem says that the square of the length of the long side of a right-angled triangle is the sum of the squares of the other two sides. In just the same way, the square of the length of a complex number is given by the the sum of the squares of its ordinary and imaginary parts. And this is how you get the probability from the complex number given by Ψ. It is a tribute to the earliest mathematicians, that the very first mathematical theorem we know of turns out to lie at the centre of quantum physics.

When we decide to measure some feature of a system, like the position of a ball or the spin of an electron, there is a certain set of possible outcomes. Quantum theory tells us how to convert the wavefunction Ψ into a probability for each outcome, using the Pythagorean theorem. And this is all quantum theory ever predicts. Often, when we are trying to predict the behaviour of large objects, the probabilities will hugely favour one outcome. For example, when you drop a ball, quantum theory predicts it will fall with near certainty. But if you let a tiny subatomic particle go, its position will soon become more and more uncertain. In quantum theory, it is in general only large collections of particles which together behave in highly predictable ways.

On the right of the equation, there are two funny symbols, which look like tall, thin, stretched-out S's. They are called integral signs, and they tie everything together. The large one tells you to add up the contributions from every possible history of the world that ends at that particular state. For example, if we let our little particle go at one position and wanted to know how likely it was to turn up at some other position at some later time, we would consider all the possible ways it could have travelled between the two positions. It might go at a fixed speed and in a straight line. Or it could jump over to the moon and back. Each one of these possible paths contributes to the final wavefunction, Ψ. It is as if the world has this incredible ability to survey every possible route to every possible future, and all of them contribute to Ψ. The U.S. physicist Richard Feynman discovered this formulation of quantum theory, known as the "sum over histories," and it is the language in which our formula for all known physics is phrased.

What is the contribution of any one history? That is given by everything to the right of the large integral sign, $\int$. First, we see the number named e by the eighteenth-century Swiss mathematician Leonhard Euler. Its value is 2.71828...If you raise e to a power, it describes exponential growth, found in many real-life situations, from the multiplication of bacteria in a culture, to the growth of money according to compound interest, or the power of computers according to Moore's law. It even describes the expansion of the universe driven by vacuum energy.

But the use of e in the formula is cleverer than that. Euler discovered what is sometimes called "the most remarkable formula in mathematics," connecting algebra and analysis to geometry: if you raise e to a power that is imaginary—meaning it is an ordinary number times i—you get a complex number for which the sum of the squares of the ordinary and imaginary parts is one. In quantum theory, this fact ensures that the probabilities for all possible outcomes add up to one. Quantum theory therefore connects algebra, analysis, and geometry to probability, combining almost all of the major areas of mathematics into our most fundamental description of nature.

In the formula for all known physics, e is raised to a power that includes all the known laws of physics in a combination called the "action." The action is the quantity starting with the small integral sign, $\int$. That symbol means you have to add up all six terms to the right of it, over all space and all time, leading up to the moment for which you wish to know the Schrödinger wavefunction Ψ. The action is just an ordinary number, but one that is associated with any possible history of the world.

As we discussed in Chapter Two, the formulation of the laws of physics in terms of an action was developed early in the nineteenth century by the Irish mathematical physicist William Rowan Hamilton. This combination of the classical laws of physics (as represented in the action), the imaginary number i, Planck's constant h, and Euler's number e together represent the quantum world. The two stretched out S's represent its exploratory, holistic character. If only we could see inside our formula and directly experience the weird and remote quantum world without having to reduce it to a set of outcomes, each assigned a probability, we might see a whole new universe inside it.

LET US NOW WALK through the six terms in the action, which together represent all the known physical laws. In sequence, they are: the law of gravity; the three forces of particle physics; all the matter particles; the mass term for matter particles; and finally, two terms for the Higgs field.

In the first term, gravity is represented by the curvature of spacetime, R, which is a central quantity in Einstein's theory of gravity. Also appearing is G, Newton's universal constant of gravitation. This is all that remains, in fundamental physics, of Newton's original laws of motion and gravity.

In the second term, F stands for fields like those James Clerk Maxwell introduced to describe electric and magnetic forces. In our very compact notation, the term also represents the fields of the strong nuclear force, which holds atomic nuclei together, and the weak nuclear force, which governs radioactivity and the formation of the chemical elements in stars. Both are described using a generalization of Maxwell's theory developed in the 1950s by Chinese physicist Chen-Ning Yang and U.S. physicist Robert Mills. In the 1960s, U.S. physicists Sheldon Lee Glashow and Steven Weinberg and Pakistani physicist Abdus Salam unified the weak nuclear force and electromagnetism into the "electroweak" theory. In the early 1970s, Dutch physicist Gerard 't Hooft and his doctoral advisor, Martinus Veltman, demonstrated the mathematical consistency of quantum Yang-Mills theory, adding great impetus to these models. And soon after, U.S. physicists David J. Gross, H. David Politzer, and Frank Wilczek showed that the strong nuclear force could also be described by a version of Yang–Mills theory.

The third term was invented in 1928 by the English physicist Paul Dirac. In thinking about how to combine relativity with quantum mechanics, he discovered an equation that describes elementary particles like electrons. The equation turned out to also predict the existence of antimatter particles. Dirac noted that for every particle — like the electron, with a definite mass and electric charge — his equation predicted another particle, with exactly the same mass but the opposite electric charge. This stunning prediction was made in 1931; the following year, the U.S. physicist Carl D. Anderson detected the positron, the electron's antimatter partner, with the exact predicted properties.

Dirac's equation describes all the known matter particles, including electrons, muons, taons, and their neutrinos, and six different types of quarks. Each one has a corresponding antimatter particle. Both the particles and the antiparticles are quanta of a Dirac field, denoted by ψ, the lower case Greek letter *psi*. The Dirac term in the action also tells you how all these particles interact through the strong and electroweak forces and gravity.

The fourth term was introduced by the Japanese physicist Hideki Yukawa, and developed into its detailed, modern form by his compatriots Makoto Kobayashi and Toshihide Maskawa in 1973. This term connects Dirac's field ψ to the Higgs field φ, which we shall discuss momentarily. The Yukawa–Kobayashi–Maskawa term describes how all the matter particles get their masses, and it also neatly explains why antimatter particles are not quite the perfect mirror images of their matter particle counterparts.

Finally, there are two terms describing Higgs field φ, the lower case Greek letter pronounced *phi*. The Higgs field is central to the electroweak theory.

One of the key ideas in particle physics is that the force-carrier fields and matter particles, all described by Maxwell–Yang–Mills theory or Dirac's theory, come in several copies. In the early 1960s, a theoretical mechanism was discovered for creating differences between the copies, giving them different masses and charges. This is the famous Higgs mechanism. It was inspired by the theory of superconductivity, where the electromagnetic fields are squeezed out of superconductors. Philip Anderson, a famous U.S. condensed matter physicist, suggested that this mechanism might operate in the vacuum of empty space. The idea was subsequently combined with Einstein's theory of relativity by several particle theorists, including the Belgian physicists Robert Brout and François Englert and the English physicist Peter Higgs. The idea was further developed by the U.S. physicists Gerald Guralink and Carl Hagen, working with the English physicist Tom Kibble, who I was fortunate to have as one of my mentors during my Ph.D.

The Higgs mechanism lies at the heart of Glashow, Salam, and Weinberg's theory, in which the electroweak Higgs field φ is responsible for separating Maxwell's electromagnetic force out from the weak nuclear force, and fixing the basic masses and charges of the matter particles.

The last term, the Higgs potential energy, $V(\varphi)$, ensures that the Higgs field φ takes a fixed constant value in the vacuum, everywhere in space. It is this value that communicates a mass to the quanta of the force fields and to the matter particles. The Higgs field can also travel in waves — similar to electromagnetic waves in Maxwell's theory — that carry energy quanta. These quanta are called "Higgs bosons." Unlike photons, they are fleetingly short-lived, decaying quickly into matter and antimatter particles. They have just been discovered at the Large Hadron Collider in the CERN laboratory in Geneva, confirming predictions made nearly half a century ago.

Finally, the value of the Higgs potential energy, V, in the vacuum also plays a role in fixing the energy of empty space — the vacuum energy — measured recently by cosmologists.

Taken together, these terms describe what is known as the "Standard Model of Particle Physics." The quanta of force fields, like the photon and the Higgs boson, are called "force-carrier particles." Including all the different spin states, there are thirty different force-carrier particles in total, including photons (quanta of the electromagnetic field), W and Z bosons (quanta of the weak nuclear force field), gluons (quanta of the strong force), gravitons (quanta of the gravitational field), and Higgs bosons (quanta of the Higgs field). The matter particles are all described by Dirac fields. Including all their spin and antiparticle states, there are a total of ninety different matter particles. So, in a sense, Dirac's equation describes three-quarters of known physics.

When I was starting out as a graduate student I found the existence of this one-line formula, which summarizes everything we know about physics, hugely motivating. All you have to do is master the language and learn how to calculate, and in principle you understand at a basic level all of the laws governing every single physical process in the universe.

. . .

YOU MAY WELL WONDER how it was that physics converged on this remarkably simple unified formula. One of the most important ideas guiding its development was that of "symmetry." A symmetry of a physical system is a transformation under which the system does not change. For example, a watch ticks at exactly the same rate wherever you place it, because the laws governing the mechanism of the watch do not depend on where the watch is. We say the laws have a symmetry under moving the watch around in *space*. Similarly, the watch's working is unchanged if we rotate the watch — we say the laws have a symmetry under *rotations*. And if the watch works just the same today, or tomorrow, or yesterday, or an hour from now, we say the laws that govern it have a symmetry under shifts in *time*.

The entry of these ideas of symmetry into physics traces back to a remarkable woman named Emmy

Noether, who in 1915 discovered one of the most important results in mathematical physics. Noether showed mathematically that any system described by an action that is unchanged by shifts in time, as most familiar physical systems are, automatically has a conserved energy. Likewise, for many systems it makes no difference to the evolution of the system exactly where the system is located in space. What Noether showed in this case is that there are three conserved quantities—the three components of momentum. There is one of these for each independent direction in which you can move the system without changing it: east–west, north–south, up–down.

Ever since Newton, these quantities—energy and momentum—had been known, and found to be very useful in solving many practical problems. For example, energy can take a myriad forms: the heat energy stored in a boiling pan of water, the kinetic energy of a thrown ball, the potential energy of a ball sitting on a wall and waiting to fall, the radiation energy carried in sunlight, the chemical energy stored up in oil or gas, or the elastic energy stored in a stretched string. But as long as the system is isolated from the outside world, and as long as spacetime is not changing (which is an excellent approximation for any real experiment conducted on Earth), the total amount of energy will remain the same.

The total momentum of a system is another very useful conserved quantity, for example in describing the outcome of collisions. Similarly, Benjamin Franklin's law of electric charge conservation, that you can move charge around but never change its total amount, is another consequence of Noether's theorem.

Before Emmy Noether, no one had really understood why *any* of these quantities are conserved. What Noether realized was as simple as it was profound: the conservation laws are mathematical consequences of the symmetries of space and time and other basic ingredients in the laws of physics. Noether's idea was critical to the development of the theories of the strong, weak, and electroweak forces. For example, in electroweak theory, there is an abstract symmetry under which an electron can be turned into a neutrino, and vice versa. The Higgs field differentiates between the particles and breaks the symmetry.

Noether was an extraordinary person. Born in Germany, she faced discrimination as both a Jew and a woman. Her father was a largely self-taught mathematician. The University of Erlangen, where he lectured, did not normally admit women. But Emmy was allowed to audit classes and was eventually given permission to graduate. After struggling to complete her Ph.D. thesis (which she later, with typical modesty, dismissed as "crap") she taught for seven years at the university's Mathematical Institute, without pay.

She attended seminars at Göttingen given by some of the most famous mathematicians of the time—David Hilbert, Felix Klein, Hermann Minkowski, and Hermann Weyl—and through these interactions her great potential became evident to them. As soon as Göttingen University's restrictions on women lecturers were removed, Hilbert and Klein recruited Noether to teach there. Against great protests from other professors, she was eventually appointed — again, without pay. In 1915, shortly after her appointment, she discovered her famous theorem.

Noether's theorem not only explains the basic conserved quantities in physics, like energy, momentum, and electric charge, it goes further. It explains how Einstein's equations for general relativity are consistent even when space is expanding and energy is no longer conserved. For example, as we discussed in the previous chapter, it explains how the vacuum energy can drive the exponential expansion of the universe, creating more and more energy without violating any physical laws.

When Noether gave her explanation for conserved quantities and more general situations involving gravity, she did so in the context of classical physics and its formulation in terms of Hamilton's action principle. Half a century later, it was realized — by the Irish physicist John Bell, along with U.S. physicists Steven Adler and Roman Jackiw — that quantum effects, included in Feynman's sum over histories, could spoil the conservation laws that Noether's argument predicted.

Nevertheless, it turns out that for the pattern of particles and forces seen in nature, there is a very delicate balance (known technically as "anomaly cancellation") that allows Noether's conservation laws to survive. This is another indication of the tremendous unity of fundamental physics: the whole works only because of all of the parts. If you tried, for example, to remove the electron, muon, and tanon and their neutrinos from physics and kept only the quarks, then Noether's symmetries and conserved quantities would be ruined and the theory would be mathematically inconsistent. This idea, that Noether's laws must be preserved within any consistent unified theory, has been a key guiding principle in the development of unified theories, including string theory, in the late twentieth century.

Noether's dedicated mentorship of students was exemplary — she supervised a total of sixteen Ph.D. students through a very difficult time in Germany's history. When Hitler came to power in 1933, Jews became targets. Noether was dismissed from Göttingen, as was her colleague Max Born. The great mathematical physicist Hermann Weyl, also working there, wrote later: "Emmy Noether — her courage, her frankness, her unconcern about her own fate, her conciliatory spirit — was, in the midst of all the hatred and meanness, despair and sorrow surrounding us, a moral solace."[2]

Eventually, Noether fled to the United States, where she became a professor at Bryn Mawr College, a women's college known particularly as a safe haven for Jewish women. Sadly, at the age of fifty-three she died of complications relating to an ovarian cyst.

In a letter to the *New York Times*, Albert Einstein wrote: "In the judgement of the most competent living mathematicians, Fräulein Noether was the most significant creative mathematical genius thus far produced since the higher education of women began. In the realm of algebra, in which the most gifted mathematicians have been busy for centuries, she discovered methods that have proved of enormous importance in the development of the present-day younger generation of mathematicians."[3] Emmy Noether was a pure soul whose mathematical discoveries opened many paths in physics and continue to exert great influence.

PAUL DIRAC WAS ANOTHER mathematical prodigy from a humble background. His discoveries laid the basis for the formula for all known physics. A master of quantum theory, he was largely responsible for its current formulation. When asked what his greatest discovery had been, he said he thought it was his "bra-ket" notation. This is a mathematical device which he introduced into quantum theory to represent the different possible states of a system. The initial state is called a "ket" and the final state a "bra." It's funny that someone who discovered the equation for three-quarters of all the known particles, who predicted antimatter, and who made countless other path-breaking discoveries would rate them all below a simple matter of notation. As with many other Dirac stories, one can't help thinking: he can't have been serious! But no one could tell.

A recent biography called Dirac "the strangest man." He was born in Bristol, England, to a family of modest means. His Swiss father, Charles Dirac, was a French teacher and a strict disciplinarian. Paul led an unhappy, isolated childhood, although he was always his father's favourite. He was fortunate to attend one of the best non-fee-paying schools for science and maths in England — Merchant Venturers' Technical College in Bristol, where his father taught.

At school, it became clear that Paul had exceptional mathematical talent, and he went on to Cambridge to study engineering. In spite of graduating with a first-class degree, he could not find a job in the postwar economic climate. Engineering's loss was physics' gain: Dirac returned to Bristol University to take a second bachelor's degree, this time in mathematics. And then, in 1923, at the ripe old age of twenty-one, he returned to St. John's College in Cambridge to work towards a Ph.D. in general relativity and quantum theory.

Over the next few years, this shy, notoriously quiet young man — almost invisible, according to some — made a series of astonishing breakthroughs. His work combined deep sophistication with elegant simplicity and clarity. For his Ph.D., he developed a general theory of transformations that allowed him to present quantum theory in its most elegant form, still used today. At the age of twenty-six, he discovered the Dirac equation by combining relativity and quantum theory to describe the electron. The equation explained the electron's spin and predicted the existence of the electron's antiparticle, the positron. Positrons are now used every day in medical PET (positron emission tomography) scans, used to track the location of biological molecules introduced into the body.

When someone asked Dirac, "How did you find the Dirac equation?" he is said to have answered with: "I found it beautiful." As often, he seemed to take pleasure in being deliberately literal and in using as few words as he could. His insistence on building physics on principled mathematical foundations was legendary. In spite of having initiated the theory of quantum electrodynamics, which was hugely successful, he was never satisfied with it. The theory has infinities, created by quantum fluctuations in the vacuum. Other physicists, including Richard Feynman, Julian Schwinger, and Sin-Itiro Tomonaga, found ways to control the infinities through a calculational technique known as "renormalization." The technique produced many accurate predictions, but Dirac never trusted it because he felt that serious mathematical difficulties were being swept under the rug. He went so far as to say that all of the highly successful predictions of the theory were probably "flukes."

Dirac also played a seminal role in anticipating the form of our formula for all known physics. For it was Dirac who saw the connection between Hamilton's powerful action formalism for classical physics and the new quantum theory. He realized how to go from a classical theory to its quantum version, and how quantum physics extended the classical view of the world. In his famous textbook on quantum theory, written in 1930 and based upon this deep understanding, he outlined the relationship between the Schrödinger wavefunction, Hamilton's action, and Planck's action quantum. Nobody followed up this insightful remark until 1946, when Dirac's comment inspired Feynman, who made the relation precise.

Dirac continued throughout his life to initiate surprising and original lines of research. He discussed the existence of magnetic monopoles and initiated the first serious attempt to quantize gravity. Although he was one of quantum theory's founders, Dirac clearly loved the geometrical Einsteinian view of physics. In some ways, one can view Dirac as a brilliant technician, jumping off in directions that had been inspired by Einstein's more philosophical work.

In his *Scientific American* article in May 1963, titled "The Evolution of the Physicists' Picture of Nature," he says, "Quantum theory has taught us that we have to take the process of observation into account, and observations usually require us to bring in the three-dimensional sections of the four-dimensional picture of the universe." What he meant by this was that in order to calculate and interpret the predictions of quantum theory, one often has to separate time from space. Dirac thought that Einstein's spacetime picture and the split into space and time created by an observer were fundamental and unlikely to change. But he suspected that quantum theory and Heisenberg's uncertainty relations would probably not survive in their current form. "Of course, there will not be a return to the determinism of classical physical theory. Evolution does not go backward," he says. "There will have to be some new development that is quite unexpected, that we cannot make

a guess about, which will take us still further from classical ideas."

Many physicists regarded the unworldly Dirac with
awe. Niels Bohr said, "Of all physicists, Dirac has the
purest soul." And "Dirac did not have a trivial bone in
his body."[4] The great U.S. physicist John Wheeler said,
simply, "Dirac casts no penumbra."[5]

I met Dirac twice, both times at summer schools for
graduate students. At the first, in Italy, he gave a one-
hour lecture on why physics would never make any progress until we understood how to predict the exact value
of the electric charge carried by an electron. During the
school, there was an evening event called "The Glorious
Days of Physics," to which many of the great physicists
from earlier days had been invited. They did their best to
inspire and encourage us students with stories of staying
up all night poring over difficult problems. But Dirac, the
most distinguished of them all, just stood up and said,
"The 1920s really were the glorious days of physics, and
they will never come again." That was all he said—not
exactly what we wanted to hear!

At the second summer school where I met him, in
Edinburgh, another lecturer was excitedly explaining
supersymmetry—a proposed symmetry between the
forces and matter particles. He looked to Dirac for support, repeating Dirac's well-known maxim that mathematical beauty was the single most important guiding
principle in physics. But again Dirac rained on the parade,
saying, "What people never quote is the second part of

my statement, which is that if there is no experimental evidence for a beautiful idea after five years, you should abandon it." I think he was, at least in part, just teasing us. In his *Scientific American* article[6] he gave no such caveat. Writing about Schrödinger's discovery of his wave equation, motivated far more by theoretical than experimental arguments, Dirac said, "I believe there is a moral to this story, namely that it is more important to have beauty in one's equations than to have them fit experiment."

Dirac ended his article by advocating the exploration of interesting mathematics as one way for us to discover new physical principles: "It seems to be one of the fundamental features of nature that fundamental physical laws are described in terms of a mathematical theory of great beauty and power, needing quite a high standard of mathematics for one to understand it. You may wonder: why is nature constructed along these lines? One can only answer that our present knowledge seems to show that nature is so constructed. We simply have to accept it. One could perhaps describe the situation by saying that God is a mathematician of a very high order and He used very advanced mathematics in constructing the universe. Our feeble attempts at mathematics enable us to understand a bit of the universe, and as we proceed to develop higher and higher mathematics we can hope to understand the universe better."

Dirac's God was, I believe, the same one that Einstein or the ancient Greeks would have recognized: the God that is nature and the universe, and whose works epitomize the very best in rationality, order, and beauty. There is no higher compliment that Dirac can pay than to call God "a mathematician of a very high order." Note, even here, Dirac's understatement.

Perhaps because of his shy, taciturn nature and his technical focus, Dirac is far less famous than other twentieth-century physics icons. But his uniquely logical, mathematical mind allowed him to articulate quantum theory's underlying principles more clearly than anyone else. After the 1930s, he initiated a number of research directions far ahead of his time. Above all, his uncompromising insistence on simplicity and absolute intellectual honesty continues to inspire attempts to improve on the formula he did so much to found.

. . .

AS BEAUTIFUL AS IT is, we know our magic formula isn't a final description of nature. It includes neither dark matter nor the tiny masses of neutrinos, both of which we know to exist. However, it is easy to conceive of amendments to the formula that would correct these omissions. More experimental evidence is needed to tell us exactly which one of them to include.

The second reason that the formula is unlikely to be the last word is an aesthetic one: as it stands it is only superficially "unified." Buried in its compact notation are no less than nineteen adjustable parameters, fitted to experimental measurements.

The formula also suffers from a profound logical flaw. Starting in the 1950s, it was realized that in theories like quantum electrodynamics or electroweak theory, vacuum fluctuations can alter the effective charges on matter particles at very short distances, in such a way as to make theories inconsistent. Technically, the problem is known as the "Landau ghost," after the Russian physicist Lev D. Landau.

The problem was circumvented by "grand unified" theories when they were introduced in the 1970s. The basic idea was to combine Glashow, Salam, and Weinberg's electroweak force and Gross, Politzer, and Wilczek's strong nuclear force into a single, grand unified force. At the same time, all the known matter particles would be combined into a single, grand unified particle. There would be new Higgs fields to separate out the strong and electroweak forces and distinguish the different matter particles from one another. These theories overcame Landau's problem, and for a while they seemed to be mathematically consistent descriptions of all the known forces except gravity.

Further encouragement came from calculations that extrapolated the strong force and the two electroweak forces to very short distances. All three seemed to unify nicely at a minuscule scale of around a ten-thousand-trillionth the size of a proton, the atomic nucleus of hydrogen. For a while, from aesthetic and logical grounds as well as hints from the data, this idea of grand unification seemed very appealing. The devil is in the details, however. There turned out to be a great number of different possible grand unified theories, each involving different fields and symmetries. There are a large number of adjustable parameters that have to be fitted to the observed data. The early hints of unification at very tiny scales faded as measurements improved: unification could only be achieved by adding even more fields. Instead of making physics simpler and more beautiful, grand unified theories have, so far, turned out to make it more complex and arbitrary.

A second reason to question grand unification is that its most striking predictions have not been confirmed. If at the most fundamental level there is only one type of particle, and if all of the differences between the particles we see are due to Higgs fields in the vacuum, then there should be physical processes allowing any one kind of particle to turn into any other kind of particle by burrowing quantum-mechanically through the grand unified Higgs field. One of the most dramatic such processes is proton decay, which would cause the proton, one of the basic constituents of atomic nuclei, to decay into lighter particles. If the prediction is correct, then all atoms will disappear, albeit at an extremely slow rate. For many years, researchers have searched for signals of this process in very large tanks of very clean water, observed with highly sensitive light detectors capable of detecting the process of nuclear decay, but so far without success.

But the strongest reason to doubt grand unification is that it ignores the force of gravity. At a scale not too far below the grand unified scale — about a thousand times smaller — we reach the Planck scale, a ten-million-trillionth the size of a proton, where the vacuum fluctuations start to wreak havoc with Einstein's theory of gravity. As we go to shorter wavelengths, the quantum fluctuations become increasingly wild, causing spacetime to become so curved and distorted that we cannot calculate anything. As beautiful as it is, we believe Einstein's theory, as included in the formula, to be only a stand-in. We need new mathematical principles to understand how spacetime works at very short distances.

At the far right of the formula, the Higgs potential energy, V, also poses a conundrum. Somehow, there is an extremely fine balance in the universe between the contribution from V and the contributions from vacuum fluctuations, a fine balance that results in a minuscule positive vacuum energy. We do not understand how this balance occurs. We can get the formula to agree with observations by adjusting V to 120 decimal places. It works, but it gives us no sense that we know what we are doing.

To summarize: all the physics we know can be combined into a formula that, at a certain level, demonstrates how powerful and connected the basic principles are. The formula explains many things with exquisite precision. But in addition to its rather arbitrary-looking pattern of particles and forces, and its breakdown at extremely short distances due to quantum fluctuations, it has two glaring, overwhelming failures. So far, it fails to make sense of the universe's singular beginning and its strange, vacuous future.

In practice, physicists seldom use the complete formula. Most of physics is based on approximations, on knowing which parts of the formula to ignore and how to simplify the parts you keep. Nevertheless, many predictions based on the formula have been worked out and verified, sometimes with extreme precision. For example, an electron has spin, and this causes it to behave in some respects like a tiny bar magnet. The relevant parts of the formula allow you to calculate the strength of this little bar magnet to a precision of about one part in a trillion. And the calculations agree with experiment.

For anything even slightly more complicated — like the structure of complex molecules, or the properties of glass or aluminum, or the flow of water — we are unable to work out all of the predictions because we are not good enough at doing the math, even though we believe the formula contains within it all the right answers. In the future, as I will describe in the next chapter, the development of quantum computers may completely transform our ability to calculate and to translate the magic formula directly into predictions for many processes far beyond the reach of computation today.

HOW SHALL THE BASIC problems of the indescribable beginning and the puzzling future of the universe be resolved? The most popular candidate for replacing our formula for all known physics is a radically different framework known as string theory, as mentioned in the previous chapter. String theory was discovered more or less by accident in 1968, by a young Italian post-doctoral researcher named Gabriele Veneziano, working at the European Organization for Nuclear Research (CERN) in Geneva. Veneziano wasn't looking for a unified theory; he was trying to fit experimental data on nuclear collisions. By chance, he came across a very interesting mathematical formula invented by the eighteenth-century Swiss mathematician Leonhard Euler — the very same Euler whose mathematical discoveries are central to the formula for all known physics.

Veneziano found he could use another formula of Euler's, called "Euler's beta function," to describe the collisions of nuclear particles in an entirely new way. Veneziano's calculations caused great excitement at the time, and even more so when it was realized that they were describing the particles as if they were little quantum pieces of string, an entirely different picture from that of quantum fields. Ultimately, the idea failed as a description of nuclear physics. It was superseded by the field theories of the strong and weak nuclear forces, and by the understanding that nuclear particles are complicated agglomerations of fields held together by vacuum fluctuations. But the mathematics of string theory turned out to be very rich and interesting, and during the early 1970s, it was developed rapidly.

String is envisaged as a form of perfect elastic. It can

exist as pieces with two ends or in the form of closed loops. Waves travel along it at the speed of light. And pieces of string can vibrate and spin in a myriad ways. One of string theory's most attractive features is that just one entity — string — describes an infinite variety of objects. So string theory is a highly unified theory.

In 1974, French physicist Joël Scherk and U.S. physicist John Schwarz realized that a closed loop of string, also spinning end over end, behaved like a graviton, the basic quantum of Einstein's theory of gravity. And so it turned out that string theory automatically provided a theory of quantum gravity, a totally unexpected discovery. Even more surprising, string theory seems to be free of the infinities that plague more conventional approaches to quantum gravity. In the mid-1980s, just as hopes for a grand unified theory were fading, string theory came along as the next candidate for a theory of everything.

As I discussed in Chapter Three, one of string theory's features is that it requires the existence of extra dimensions of space. In addition to the familiar three dimensions of space — length, height, and breadth — the simplest string theories require six more space dimensions, and M-theory, which I also described, requires one more, bringing the number of extra dimensions to seven. The six extra dimensions of string theory can be curled up in a little ball, so small that we would not notice them in today's universe. And the seventh dimension of M-theory is even more interesting. It takes the form of a gap between two three-dimensional worlds. This picture was the basis for the cyclic model of cosmology that I explained in the previous chapter.

Although there were great expectations that string theory would solve the problem of the unification of forces, these hopes have also dimmed. The main problem is that, like grand unified theories, string theory is itself too arbitrary. For example, it turns out to be possible to curl up the six or seven extra dimensions in an almost infinite number of ways. Each one would lead to a three-dimensional world with a different pattern of particles and forces. Most of these models are hopelessly unrealistic. Still, many researchers hope that by scanning through this "landscape" of possible string theory universes, they may find the right one. Some even believe that every one of these landscapes of universes must be realized somewhere in the actual universe, although only one of them would be visible to us. This picture, called the "inflationary multiverse," has to be one of the most extravagant proposals in the history of science.

From my own point of view, none of these string theory universes is yet remotely realistic, because string theory has so far proven incapable of describing the initial singularity, the problem I outlined in the previous chapter. The string theory landscape, so far as it is currently understood, consists of a set of empty universe models. But there are serious grounds for doubt as to whether these empty models can actually be used to describe expanding universes full of matter and radiation, like ours.

Rather than speculate about a "multiverse" of possible universes, I prefer to focus on the one we know exists, and try to understand the principles that might resolve its major puzzles: the singularity and the distant future. String theory is a powerful theoretical tool that has already provided completely new insights into quantum gravity. But there is some way to go before it is ready to convincingly describe our universe.

THE SITUATION IN WHICH string theory finds itself is in many ways a reflection of how fundamental physics developed over the course of the twentieth century. In the early part of the century came the great ideas of quantum physics, spacetime, and general relativity. There was great philosophical richness in the debates over these matters, with much fewer publications and conferences than today, and a greater premium on originality. In the late 1920s, with the establishment of quantum theory and the quantum theory of fields, attention turned to more technical questions. Physicists focused on applications and became more like technicians. They extended the reach of physics to extremely small and large distances without having to add any revolutionary new ideas.

Physics became a fertile source of new technologies — everything from nuclear power to radar and lasers, to transistors, LEDs, integrated circuits and other devices, to medical X-ray, PET, and NMR scans, and even superconducting trains. Particle accelerators probing very high energies made spectacular discoveries — of quarks, of the strong and electroweak forces, and most recently of the Higgs boson. Cosmology became a true observational science, and dedicated satellites mapped the whole universe with exquisite precision. Physics seemed to be steaming towards a final answer, towards a theory of everything.

From the 1980s on, waves of enthusiasm swept the field only to die out nearly as quickly as they arose. Publications and citations soared and conferences multiplied, but genuinely new ideas were few and far between. The mainstreaming of grand unification and string theory, and the sheer pressure it created to force a realistic model out of incomplete theoretical frameworks so far has been dissatisfying.

The development in physics is, I think, a kind of ultraviolet catastrophe, like the one Planck and Einstein discovered in classical physics at the start of the twentieth century. They are consequences of mechanical ways of thinking. I believe it is time for physics to step away from contrived models, whether artificial mathematical constructs or *ad hoc* fits to the data, and search for new unifying principles. We need to better appreciate the magic we have discovered, and all of its limitations, and find new ways to see into and beyond it.

Every term in our formula required a giant leap of the imagination — from Einstein's description of gravity, to Dirac's description of the electron and other particles, to Feynman's formulation of quantum mechanics as a sum over all possible histories. We need to foster opportunities for similar leaps to be made. We need to create a culture where the pursuit of deep questions is encouraged and enabled: where the philosophical richness and depth of an Einstein or Bohr combine with the technical brilliance of a Heisenberg or Dirac.

As I have emphasized, some of the greatest contributions to physics were made by people from very ordinary backgrounds who, more or less through chance, came to work on fundamental problems. What they had in common was the boldness to follow logical ideas to their conclusion, to see connections everyone else had missed, to explore unknown territories, and to play with entirely new ideas. And this boldness produced leaps of understanding way beyond everyday experience, way beyond our circumstances and our history, leaps which we can all share.

. . .

WHEN CHILDREN GO TO school, we teach them algebra and geometry, physics according to Newton's laws, and so on, but as far as I know, nobody says anything about the fact that physics has discovered a blueprint of the universe. Although the formula takes many years of study to fully understand and appreciate, I believe it is inspirational to realize how far we have come towards combining the fundamental laws that govern the universe.

In its harmonious and holistic nature, the formula is, I believe, a remarkable icon. All too often, our society

today is driven by selfish behaviour and rigid agendas—
on the one hand by people and groups pursuing their
own short-term interests, and on the other by appeals to
preconceived systems that are supposed to solve all our
problems. But almost all of the traditional prescriptions
have failed in the past, and they are all prone to being
implemented in inhuman ways. It seems to me that as we
enter a period of exploding human demand and increasingly limited resources, we need to look for more intelligent ways to behave.

The formula suggests principles that might be more
useful. In finding the right path for society, perhaps
we need to consider all paths. Just as quantum theory
explores all options and makes choices according to
some measure of the "benefit," we need to run our societies more creatively and responsively, based on a greater
awareness of the whole. The world is not a machine that
we can set in some perfect state or system and then forget about. Nor can we rely on selfish or dogmatic agendas as the drivers of progress. Instead, we need to take an
informed view of the available options and be far-sighted
enough to choose the best among them.

It is all too easy to define ourselves by our language,
nationality, religion, gender, politics, or culture. Certainly we should celebrate and draw strength from our
diversity. But as our means of communications amplify,
these differences can create confusion, misunderstand-
ing, and tension. We need more sources of commonality,
and our most basic understanding of the universe, the

place we all share, serves as an example. It transcends all our differences and is by far the most reliable and cross-cultural description of the world we have. There is only one Dirac or Einstein or Maxwell equation, and each of these is so simple, accurate, and powerful that people from any and all backgrounds find it utterly compelling. Even the failures of our formula are something we can all agree about.

And that, I think, will be the key to future scientific breakthroughs. If you look at the people who made the most important contributions in the past, many were among the first members of their societies to get involved in serious science. Many faced discrimination and prejudice. In overcoming these obstacles they had a point to prove, which encouraged them to question traditional thinking. As we saw in Chapter Two, many of the most prominent twentieth-century physicists were Jewish, yet until the mid-nineteenth century Jews had been deliberately excluded from science and technical subjects at many universities across Europe. When they finally gained access, they were hugely motivated to disprove their doubters, to show that Jews could do every bit as well as anyone else. Einstein, Bohr, Born, and Noether were part of an influx of new talent that completely revolutionized physics in the early twentieth century.

Which brings me back to the question of unification, both of people across the planet and of our understanding of the world. The search for a superunified theory is an extremely ambitious goal. *A priori*, it would seem to be hopeless: we are tiny, feeble creatures dwarfed by the universe around us. Our only tools are our minds and our ingenuity. But these have enabled us to come amazingly far. If we think of the world today, with seven billion minds, many in emerging economies and societies, it is clear that there is a potential gold mine of talent. What is needed is to open avenues for gifted young people to enter and contribute to science, no matter what their background. If opportunities are opened, we can anticipate waves of motivated, original young people capable of transformative discoveries.

Who are we, in the end? As far as we know, we represent something very rare in the universe — the organization of matter and energy into living, conscious beings. We have learned a great deal about our origins — about how the universe emerged from the singularity filled with a hot plasma; how the chemical elements were created in the big bang and stars and supernovae; how gravity and dark matter clumped molecules and atoms together into galaxies, stars, and planets; how Earth cooled and allowed lakes and oceans to condense, creating a primordial soup within which the first life arose. We do not know exactly how life started, but once the first self-regulating, self-replicating organisms formed, containing the DNA-protein machinery of life, reproduction, competition, and natural selection drove the evolution of more and more complex living organisms. We humans stand now on the threshold of a new phase of evolution, in which technology will play as much of a role as biology.

Great mysteries remain. Why did the universe emerge from the big bang with a set of physical laws that gave rise to heavy elements and allowed complex chemistry? Why did these laws allow for planets to form around stars, with water, organic molecules, an atmosphere, and the other requirements for life? Why did the DNA-protein machinery, developed and selected for in the evolution of primitive single-cell organisms, turn out to be able to code for complex creatures, like ourselves? How and why did consciousness emerge?

At every stage in the history of the universe, there was the potential for vastly more than what had been required to reach that stage. Today, this is more true than ever. Our understanding of the universe has grown faster than anyone could have imagined a century ago, way beyond anything that could be explained in terms of past evolutionary advantage. We cannot know what new technologies we will create, but if the past is any guide, they will be extraordinary. Commercial space travel is about to become a reality. Quantum computers are on the horizon, and they may completely transform our experience of the world. Are all these capabilities simply accidental? Or are we actually the door-openers to the future? Might we be the means for the universe to gain a consciousness of itself?

FIVE

THE OPPORTUNITY OF ALL TIME

Whence things have their origin,
Thence also their destruction happens,
As is the order of things;
For they execute the sentence upon one another
— The condemnation for the crime —
In conformity with the ordinance of Time.
— Anaximander[1]

ANAXIMANDER'S QUOTE MIGHT HAVE been meant for today: our world is changing rapidly, with our future in the balance.

Our global population has grown to seven billion and continues to rise. We are eating away at our supplies of energy, water, fertile land, minerals. We are spoiling our environment and driving species extinct every day. We are caught up in financial and political crises entirely of our own creation. Sometimes it feels as if the technological progress on which we have built our lives and our societies is just leading us towards disaster. There is an overwhelming sense that we are running out of time.

Our personal capabilities have never been greater. Many of us can now communicate instantly with collaborators, friends, and family around the globe. This ability has powered new democratic movements, like those of the Arab Spring, and has allowed the assembly of great stores of collectively curated information, like Wikipedia. It is driving global scientific collaborations and opening online access to quality educational materials and lectures to people everywhere.

But the internet, with all of its attractions, is also profoundly dehumanizing. Increasingly we are glued to our computers and smartphones, building our social and professional lives around email, social media, blogs, or tweets. Overload of digital information turns us into automata, workaholics, passive consumers. Its harsh physical form stresses us and creates a mismatch between our own human nature and the manner in which we are being forced to communicate. Our analog nature is being compressed into a digital stream. Not so surprising then that, as the comedian Louis C. K. recently put it, "Everything is amazing right now and nobody's happy."[2]

Massive economic shifts are also taking place. Past political paradigms are becoming irrelevant, with Western governments intervening to shore up their financial systems, and China overseeing the world's greatest market-driven economic boom. Information is the new oil, and knowledge-based companies like Google, Amazon, and Facebook are replacing manufacturing industries in many developed Western countries. Instead of the old worker–owner division, Western society is developing new fractures: between an economically active elite and a marginalized remainder.

Short-term thinking is endemic, as is natural when things are moving fast. It is as if we are driving a speeding car through a fog, swerving to avoid potholes, roadblocks, or oncoming vehicles, anxiously anticipating the dangers with no power to predict them. Politicians tend to think no further than the next election, scientists no further than the next grant.

In this chapter, I want to talk about the future of this world of ours. The coming century will see our lives, and those of our children, transformed. What happens will depend on the decisions we take and on the discoveries we make. I won't make any forecasts. Nor will I try to outline a plan for our survival. That is a pragmatic task requiring the skills and dedication of many people.

Instead, I want to try to step back from all the anxieties and the immediate issues of today and address something more basic and long-term, namely our own human character: how our ideas regarding our place in the universe may develop, and how our very nature may change. Speaking about the future makes us nervous. Einstein said, "I never think of the future; it comes soon enough." We do not really know who we are or what we are capable of. I feel like a diver standing on the edge of a tall cliff, looking over the precipice and peering through the fog below. Is there a beautiful, cool ocean waiting for me, or only jagged rocks? I don't know. Nevertheless, I will take the plunge.

As I will explain, scientific advances we can now envisage may bring us, as living, self-conscious beings, much closer to physical reality. The separation of our ideas from our nature, of science from society, of our intellect from our feelings, and of ourselves from the universe may diminish. Not only might we see the universe more clearly, but we may come to know it more deeply. And in time, that knowledge will change who we are. This is an extraordinary prospect, which I hope will encourage us to see a more inspirational future.

THINKING ABOUT THE UNIVERSE might seem like escapism, or a luxury: how will it solve the problem of world hunger, or carbon emissions, or the national debt? But throughout history, from Anaximander and Pythagoras to Galileo and Newton, the universe has been an endless source of wonder, inspiring us to rise above our current circumstances and see what lies beyond. That basic urge continues today, driving the creation of the most powerful ever microscope — the Large Hadron Collider — and the most powerful ever telescope — the Planck satellite. It has resulted in a working mathematical model of all the forces and particles in nature, tested to precision from length scales well below the size of an atomic nucleus up to the entire visible universe. We understand the broad features of the evolution of the cosmos, from its first microseconds up to the spectacular present, where we see hundreds of billions of galaxies stretching across space. The Higgs particle — a manifestation of the mechanism through which matter particles and forces acquire their distinctive characters — has just been discovered, one of physics' crowning achievements.

Discoveries as basic as this one can take a long time for their full impact to be felt. But the more basic they are, the more profound the impact. Quantum physics was formulated in the 1920s, but it was not until the 1960s that its implications for the nature of our reality began to be more fully appreciated. We think of and discuss the world as if it were an arena filled with definite things, whose state changes from one moment to the next. This is the picture of the classical universe, as developed by Newton, Maxwell, and Einstein, evolving according to deterministic physical laws.

Quantum theory makes predictions that are inconsistent with this picture, and experiment shows them to be right. According to quantum theory, the world is constantly exploring all of its possible classical states all of the time, and is only appearing to us as any one of them with some probability. The conceptual machinery that underlies this view of quantum reality involves strange mathematical concepts like the square root of minus one, for which we have little intuition. And only now are the technological implications of these basic discoveries becoming apparent.

At the same time, fundamental research continues to identify new avenues for expanding the boundaries of our knowledge. As successful and far-reaching as our modern picture of the universe is, our description completely fails at the critical event — the big bang singularity — from which everything we now see around us emerged. Our current understanding likewise offers little explanation for the universe's strange future. The energy of empty space — the vacuum energy, which is itself controlled by quantum effects — has taken over as the dominant form of energy in the universe. In the coming tens of billions of years, its repulsive gravitational force will speed up the expansion of the universe and carry all the galaxies we now see out of our view. As Anaximander said, our world is transitory, and the physical forces that enabled its emergence are now in the process of taking it away.

The theories of the twentieth century are struggling to tackle these problems — of the emergence of the universe and of its ultimate fate. String theory is the leading contender for a "theory of everything," possessing exciting mathematical properties that suggest it might include every known force and particle. But string theory comes along with tiny extra dimensions of space, so small they are invisible, whose form fixes the pattern of forces and particles we should see. Unfortunately, the theory does not make any definite prediction for the form of the extra dimensions, and with our present understanding, the number of possible configurations seems almost uncountable. For each one of these configurations, the universe consisting of the visible dimensions and the particles and forces within them would appear very different. String theory therefore seems to predict a "multiverse" rather than a universe. Instead of being a theory of everything, it is more like a theory of anything.

String theory's lack of a definite prediction for the vacuum energy, combined with the puzzling observation that the vacuum energy takes a tiny positive value, has encouraged many scientists to embrace what seems to many of us like an unscientific explanation: that every one of these universes is possible, but the one we find ourselves in is the only one that actually allows us to exist. Sadly, this idea is at best a rationalization. It is hard to imagine a less elegant or convincing explanation of our own beautiful world than to invent a near-infinite number of unobservable worlds and to say that, for some reason we cannot understand or quantify, ours was "chosen" to exist from among them.

Most string theorists have likewise avoided the problem of the big bang singularity, although every one of their hypothesized worlds possesses such a starting point. Typically, they are content to assume the universe sprang into existence in one of the plethora of allowed forms, just after the singularity, and to discuss its evolution from there. So indeed, from the most widely accepted viewpoints, the beginning and the end of the universe seem to be brick walls beyond which physics cannot go.

The puzzles of the beginning and the future of the universe are, in my view, the critical clues which may help us rise above current paradigms and find a better vantage point. As I discussed in Chapter Three, we do have ways of conceptually taking the universe into the quantum domain, and these now suggest a very different picture, in which we may traverse the big bang singularity to a universe before it and likewise pass beyond our vacuous future into the next big bang to come. If this suggestion is correct, the implication is that there was no beginning of time nor will there be an end: the universe is eternal, into the past and into the future.

. . .

OUR SOCIETY HAS REACHED a critical moment. Our capacity to access information has grown to the point where we are in danger of overwhelming our capabilities to process it. The exponential growth in the power of our computers and networks, while opening vast opportunities, is outpacing our human abilities and altering our forms of communication in ways that alienate us from each other. We are being deluged with information through electrical signals and radio waves, reduced to a digital, super-literal form that can be reproduced and redistributed at almost no cost. The technology makes no distinction between value and junk. The abundance and availability of free digital information is dazzling and distracting. It removes us from our own nature as complex, unpredictable, passionate people.

The "ultraviolet catastrophe" that physics encountered at the end of the nineteenth century serves as a metaphor for physics today, as I have already suggested, and also for our broader crisis. Maxwell's theory of electromagnetic radiation and light was a triumph, being the most beautiful and powerful application of mathematics to describing reality. Yet it implied that there were waves of all wavelengths, from zero to infinity. In any realistic context, where heat and electromagnetic energy are constantly being exchanged between objects, this feature of Maxwell's theory leads to a disaster. Any hot object, or any electron in orbit around an atom, can radiate electromagnetic waves at an unlimited rate, leading to a disastrous instability of the world.

Planck was forced to tackle this problem by taking a step back from a literal, classical world as envisaged by Newton, Maxwell, and Einstein. Ultimately, we had to give up the idea of a definite reality comprising a geometrical arena — spacetime — inhabited by entities in the form of particles and waves. We had to give up any notion of being able to picture things as they really are, or of being able (even in principle) to measure and predict everything there is to know. These ideas had to be replaced with a more abstract, all-encompassing theory, which reduced our capacity to "know" or "visualize" reality, while giving us a powerful new means of describing and predicting nature.

In the same way, I believe we now need to step back from the overwhelming nature of our "digital age." One can already see a tendency among many people to "surf" across the ocean of information on the internet. This behaviour seems to replace a desire for a deeper or more rounded understanding of anything. Mastery seems unfeasible in a world awash with information. However, higher level thinking is needed now more than ever. We need to develop more refined skills of awareness and judgement to help us filter, select, and identify opportunities. Collaboration will increasingly be the name of the game as people around the world work, share ideas, write books, and even construct mathematical proofs together.

Viewed in this light, our modes of education at school and university seem terribly outmoded. Young people don't need to memorize known facts any more — they are all readily accessible on the internet. The skills they need most are to think for themselves, to choose what to learn, to develop ideas and share them with others. How to see the big picture, how to find just what they need in an ocean of knowledge, how to collaborate and how to dig deep in an entirely new direction.

It seems to me we need to create a modern version of the ancient Greek philosophers' fora or Scotland's educational system in the late eighteenth century, where the principles and foundations of knowledge were questioned and debated, and where creativity, originality, and humility before the truth were the most highly prized qualities in a student.

Our society has been shaped by physics' past discoveries to an extent that is seldom appreciated. The mechanical world of Newton led to mechanical ways of learning, as well as to the modern industrial age. We are all very aware of how the digital revolution is transforming our lives: computers are filling our schools and offices, replacing factory workers, miners, and farmers. They are changing the way we work, learn, live, and think. Where did this new technology come from? It came, once again, from our capacity to understand, invent, and create: from the Universe Within.

THE STORY OF HOW physics created the information age begins at the turn of the twentieth century, when electricity was becoming the lifeblood of modern society. There was galloping demand to move current around—quickly, predictably, safely—in light bulbs, radios, telegraphs, and telephones. Joseph John (J. J.) Thomson's discovery of the electron in 1897 had explained the nature of electricity and launched the development of vacuum tubes.

For most of the twentieth century, amplifying vacuum tubes were essential components of radios, telephone equipment, and many other electrical devices. They consist of a sealed glass tube with a metal filament inside that releases lots of electrons when it is heated up. The negatively charged electrons stream towards a positively charged metal plate at the other end of the tube, carrying the electrical current. This simple arrangement, called a "diode," allows current to flow only one way. In more complicated arrangements, one or more electrical grids are inserted between the cathode and anode. By varying the voltage on the grids, the flow of electrons can be controlled: if things are arranged carefully, tiny changes in the grid voltage result in large changes in the current. This is an amplifier: it is like controlling the flow of water from a tap. Gently twiddling the tap back and forth leads to big changes in the flow of water.

Vacuum tubes were used everywhere — in radios, in telephone and telegraph exchanges, in televisions and the first computers. However, they have many limitations. They are large and have to be warmed up. They use lots of power, and they run hot. Made of glass, they are heavy, fragile, and expensive to manufacture. They are also noisy, creating a background "hum" of electrical noise in any device using them.

In Chapter One, I described the Scottish Enlightenment and how it led to a flowering of education, literature, and science in Scotland. James Clerk Maxwell was one of the products of this period, as were the famous engineers James Watt, William Murdoch, and Thomas Telford; the mathematical physicists Peter Guthrie Tait and William Thomson (Lord Kelvin); and the writer Sir Walter Scott. Another was Alexander Graham Bell, who followed Maxwell to Edinburgh University before emigrating to Canada, where he invented the telephone in Brantford, Ontario — and in so doing, launched global telecommunications.

Bell believed in the profound importance of scientific research, and just as his company was taking off in the 1880s, he founded a research laboratory. Eventually christened Bell Labs, this evolved into the research and development wing of the U.S. telecommunications company AT&T, becoming one of the most successful physics centres of all time, with its scientists winning no fewer than seven Nobel Prizes.[3]

At Bell Labs, the scientists were given enormous freedom, with no teaching duties, and were challenged to do exceptional science. They were led by a visionary, Mervin Kelly, who framed Bell Labs as an "institute for creative technology," housing physicists, engineers, chemists, and mathematicians together and allowing them to pursue investigations "sometimes without concrete goals, for years on end."[4] Their discoveries ranged from the basic theory of information and communication and the first cellular telephones to the first detection of the radiation from the big bang; they invented lasers, computers, solar cells, CCDs, and the first quantum materials.

One of quantum theory's successes was to explain why some materials conduct electricity while others do not. A solid material consists of atoms stacked together. Each atom consists of a cloud of negatively charged electrons orbiting a positively charged nucleus. The outermost electrons are farthest from the nucleus and the least tightly bound to it — in conducting materials like metals, they are free to wander around. Like the molecules of air in a room, the free electrons bounce around continuously inside a piece of metal. If you connect a battery across the metal, the free electrons drift through it in one direction, forming an electrical current. In insulating materials, there are no free electrons, and no electrical currents can flow.

Shortly after the Second World War, Kelly formed a research group in solid state physics, under William Shockley. Their goal was to develop a cheaper alternative to vacuum tubes, using semiconductors — materials that are poor conductors of electricity. Semiconductors were already being used, for example, in "point-contact" electrical diodes, where a thin needle of metal, called a "cat's whisker," was placed in contact with a piece of semiconductor crystal (usually lead sulphide or galena). At certain special points on the surface, the contact acts like a diode, allowing current to flow only one way. Early "crystal" radio sets used these diodes to convert "amplitude modulated" AM radio signals into DC currents, which then drove a headset or earphone. In the 1930s, Bell scientists explored using crystal diodes for very high frequency telephone communications.

During the war, lots of effort had gone into purifying semiconductors like germanium and silicon, on the theory that removing impurities would reduce the electrical noise.[5] But it was eventually realized that the magic spots where the crystal diode effect works best correspond to *im*purities in the material. This was a key insight — that controlling the impurities is the secret to the fine control of electrical current.

Just after the war, Shockley had tried to build a semiconductor transistor, but had failed. When Kelly asked Shockley to lead the Solid State Physics group, he placed the theorist John Bardeen and the experimentalist Walter Brattain under his supervision. The two then attempted to develop the "point-contact" idea, using two gold contacts on a piece of germanium which had been "doped" — seeded with a very low concentration of impurities to allow charge to flow through the crystal.

They were confounded by surface effects, which they initially overcame only through the drastic step of immersing the transistor in water, hardly ideal for an electrical device. After two years' work, their breakthrough came in the "miracle month" of November–December 1947, when they wrapped a ribbon of gold foil around a plastic triangle and sliced the ribbon through one of the triangle's points. They then pushed the gold-wrapped tip into the germanium to enable a flow of current through the bulk of the semiconductor. A voltage applied to one of the two gold contacts was then found to amplify the electric current flowing from the other contact into the germanium, like a tap being twiddled to control the flow of water.[6]

Bardeen, Brattain, and Shockley shared the 1956 Nobel Prize in Physics for their discovery of the transistor, which launched the modern electronics age. Their "point contact" transistor was quickly superseded by "junction" transistors, eventually to be made from silicon. Soon after, the team split up. Bardeen left for the University of Illinois, where he later won a second

Nobel Prize. Shockley moved out to California, where he founded Shockley Semiconductor. He recruited eight talented young co-workers who, after falling out with him, left to form Fairchild and Intel, thereby launching Silicon Valley.

Transistors can control the flow of electricity intricately, accurately, and dependably. They are cheap to manufacture and have become easier and easier to miniaturize. Indeed, to date, making computers faster and more powerful has almost entirely been a matter of packing more and more transistors onto a single microprocessor chip.

For the past forty years, the number of transistors that can be packed onto a one-square-centimetre chip has doubled every two years — an effect known as Moore's law, which is the basis for the information and communication industry's explosive growth. There are now billions of transistors in a typical smartphone or computer CPU. But there are also fundamental limits, set by the size of the atom and by Heisenberg's uncertainty principle. Extrapolating Moore's law, transistors will hit these ultimate limits one or two decades from now.

In modern computers, information consists of strings of 0s and 1s stored in a pattern of electrical charges or currents or magnetized states of matter, and then processed via electrical signals according to the computer program's instructions. Typically, billions of operations are performed per second upon billions of memory elements. It is crucial to the computer's operation that the

0s and 1s are stored and changed accurately and not in unpredictable ways.

The problem is that the moving parts of a computer's memory — in particular, the electrons — are not easy to hold still. Heisenberg's uncertainty principle says that if we fix an electron's position, its velocity becomes uncertain and we cannot predict where it will move next. If we fix its velocity, and therefore the electrical current it carries, its position becomes uncertain and we don't know where it is. This problem becomes unimportant when large numbers of electrons are involved, because to operate a device one only needs the average charge or current, and for many electrons these can be predicted with great accuracy. However, when circuits get so tiny that only a few electrons are involved in any process, then their quantum, unpredictable nature becomes the main source of error, or "noise," in the computer's operations. Today's computers typically store one bit of data in about a million atoms and electrons, although scientists at IBM Labs have made a twelve-atom bit register called "atomic-scale memory."[7]

QUANTUM UNCERTAINTY IS THE modern version of the impurities in semiconductors. Initially impurities were seen as a nuisance, and large sums of money were spent trying to clean them away, before it was realized that the ability to manipulate and make use of them was the key to the development of cheap, reliable transistors. The same story is now repeating itself with "quantum uncertainty." As far as classical computers are concerned, quantum uncertainty is an unremovable source of noise, and nothing but a nuisance. But once we understand how to use quantum uncertainty instead of trying to fight it, it opens entirely new horizons.

In 1984, I was a post-doctoral researcher at the University of California, Santa Barbara. It was announced that the great Richard Feynman was going to come and give a talk about quantum computers. Feynman was one of our heroes, and this was an opportunity to see him first-hand. Feynman's talk focused on the question of whether there are ultimate limits to computation. Some scientists had speculated that each operation of a computer inevitably consumes a certain amount of energy, and that ultimately this would limit the size and power of any computer. Feynman's interest was piqued by this challenge, and he came up with a design that overcame any such limit.

There were several aspects to his argument. One was the idea of a "reversible" computer that never erased (or overwrote) anything stored in its memory. It turns out that this is enough to overcome the energy limit. The other new idea was how to perform computations in truly quantum ways. I vividly remember him waving his arms (he was a great showman), explaining how the quantum processes ran forwards and backwards and gave you just what you needed and no more.

Feynman's talk was entirely theoretical. He didn't speak at all about building such a device. Nor did he give any specific examples of what a quantum computer would be able to do that a classical computer could not. His discussion of the theory was quite basic, and most of the ingredients could be found in any modern textbook. In fact, there was really no reason why all of this couldn't have been said many decades ago. This is entirely characteristic of quantum theory: simply because it is so counterintuitive, new and unexpected implications are still being worked out today. Although he did not have any specific examples of the uses of a quantum computer, Feynman got people thinking just by raising the possibility. Gradually, more and more people started working on the idea.

In 1994, there came a "bolt from the blue." U.S. mathematician Peter Shor, working at Bell Labs (perhaps unsurprisingly!), showed mathematically that a quantum computer would be able to find the prime factors of large numbers much faster than any known method on a classical computer. The result caused a shockwave, because the secure encryption of data (vital to the security systems of government, banks, and the internet) most commonly relies on the fact that it is very difficult to find the prime factors of large numbers. For example, if you write down a random 400-digit number (which might take you five minutes), then even with the best known algorithm and the most powerful conceivable classical computer, it would take longer than the age of the universe to discover the number's prime factors. Shor's work showed that a quantum computer could, in principle, perform the same task in a flash.

What makes a quantum computer so much more powerful than a classical one? A classical computer is an automatic information-processing machine. Information is stored in the computer's memory and then read and manipulated according to pre-specified instructions — the program — also stored in the computer's memory. The main difference between a classical and quantum computer is the way information is stored. In a classical computer, information is stored in a series of "bits," each one of which can take just two values: either 0 or 1. The number of arrangements of the bits grows exponentially with the length of the string. So whereas there are only two arrangements for a single bit, there are four for two bits, eight for three, and there are nearly a googol (one with a hundred zeros after it) ways of arranging three hundred bits. You need five bits to encode a letter of the alphabet and about two million bits to encode all of the information in a book like this. Today, a typical laptop has a memory capacity measured in gigabytes, around ten billion bits (a byte is eight bits), with each gigabyte of memory capable of storing five thousand books.

A quantum computer works in an entirely different way. Its memory is composed of qubits, short for quantum bits. Qubits are somewhat like classical bits in that when you read them out, you get either 0 or 1. However, the resemblance ends there. According to quantum theory, the typical state for a qubit is to be in a *superposition* — a state consisting of 0 and 1 at the same time. The amount of 0 or 1 in the state indicates how probable it is to obtain 0 or 1 when the qubit is read.

The fact that the state of a qubit is specified by a continuous quantity — the proportion of 0 or 1 in the state — is a clue that it can store infinitely more information than a classical bit ever can.[8] The situation gets even more interesting when you have more than one qubit and their states are *entangled*. This means that, unlike classical bits, qubits cannot be read independently: what you measure for one of them will influence what you measure for the other. For example, if two qubits are entangled, then the result you obtain when you measure one of them will completely determine the result you obtain if you measure the other. A collection of entangled qubits forms a whole that is very much greater than the sum of its parts.

Shor used these features to make prime-number factoring go fast. Classically, if you tried to find the prime factors of a large number,[9] the brute force method would be to divide it by two as many times as you could, then three, then five, and so on, and keep going until no further divisions worked. However, what Shor realized, in essence, is that a quantum computer can perform all of these operations at the same time. Because the quantum state of the qubits in the computer simultaneously encodes many different classical states, the computations can all occur "in parallel," dramatically speeding up the operation.

Shor's discovery launched a global race to build a quantum computer, using a wide range of quantum technologies: atomic and nuclear spins, the polarization states of light, the current-carrying states of superconducting rings, and many other incarnations of qubits. In recent years, the race has reached fever pitch. At the time of writing, researchers at IBM are claiming they are close to producing a "scalable" quantum computing technology.

What will this vast increase in our information-handling capabilities mean? It is striking to compare our situation today, with the vast libraries at our fingertips and far vaster ones to come, with that of the authors of the modern scientific age. In the Wren Library in Trinity College, Cambridge, Isaac Newton's personal library consists of a few hundred books occupying a single bookcase. This was quite enough to allow him to found modern physics and mathematical science. A short walk away, in the main University Library, Charles Darwin's personal library is also preserved. His entire collection of books occupies a ten-metre stretch of shelving. Again, for one of the most profound and original thinkers in the history of science, it is a minuscule collection.

Today, on your smartphone, you can access information resources vastly greater than any library. And according to Moore's law, in a couple of decades your laptop will comfortably hold every single book that has ever been written. A laptop quantum computer will seem more like Jorge Luis Borges's Library of Babel — a fantastical collection holding every possible ordering of letters and words in a book, and therefore every book that could ever be written. With a quantum library, one might instead be able to search for all possible interesting passages of text without anyone having had to compose them.

Some of the uses of quantum computers and quantum communication are easy to anticipate. Ensuring the security of information is one of them. The codes currently used to protect access to bank accounts, computer passwords, and credit card information rely on the fact that it is hard to find the prime factors of large numbers using a classical computer. However, as Peter Shor showed, quantum computers will be able to quickly find these factors, rendering current security protocols obsolete. Also, quantum information is inherently safer from attack than classical information, because it is protected by the fundamental laws of physics. Whereas reading out classical information does nothing to change it, according to quantum physics, the mere fact of observing a quantum system almost always changes its quantum state. Through this effect, eavesdropping or hacking into quantum information can be detected. Hence quantum information can be made invulnerable to spying in ways that would be classically impossible.

Quantum computers may also transform our capacities to process data in parallel, and this could enable systems with great social benefit. One proposal now being considered is to install highly sensitive biochemical quantum detectors in every home. In this way, the detailed medical condition of every one of us could be continuously monitored. The data would be transmitted to banks of computers which would process and screen it for any signs of risk. The results of any medical treatment or dietary change or any other intervention would be constantly gathered. With access to such vast amounts of data and information-processing power, medicine would be revolutionized. We would all be participants in medical trials, on a scale and with an accuracy and breadth greater than anything seen before.

But by far the greatest impact quantum computers will have is likely to be on ourselves.

.   .   .

THE IDEA THAT OUR communication technologies change us was emphasized by the Canadian communications guru Marshall McLuhan. McLuhan's 1964 book, *Understanding Media: The Extensions of Man*, kicked off a wave of interest in the uses of mass media in all forms, from pop music and television to major corporations. McLuhan's writing is more poetic than analytical, but his basic insight was that the information content of all of these forms of mass media—from ads to games, cars, typewriters (remember, no PCs then!), books, telephones, newspapers, and so on—is less important than their physical form and their direct hold on our behaviour. He summed up this idea in his famous aphorism "The medium is the message." Today, watching people wander around, eyes glued to smartphones, texting or emailing, in the grip of their gadgets and nearly oblivious to their surroundings, you can see what he meant.

McLuhan's point was that media have been having this effect on us for millennia. If you think for two seconds, it is amazing, and faintly ridiculous, that the mere act of compressing, and so severely limiting, our ideas in writing — in the case of European languages, into words written in an alphabet of twenty-six letters — has proven to be such a powerful and society-dominating technology. Writing is a means of extracting ourselves from the world of our experience to focus, form, and communicate our ideas. The process of committing ourselves to texts — from the scriptures to textbooks, encyclopedias, novels, political pamphlets, laws, and contracts — and then allowing them to control our lives has had an enormous and undeniable effect on who we are. McLuhan argued that print altered our entire outlook, emphasizing our visual sense, thus influencing the fragmentation and specialization of knowledge, and fostering everything from individualism to bureaucracy to nationalistic wars, peptic ulcers, and pornography.

McLuhan saw every mass medium, whether print, photography, radio, or TV, in a similar way: as an extension of our own nervous system, dramatically altering our nature and hence our society. "We have never stopped drastically interfering with ourselves by every technology we could latch on to," he said in "The Future of Man in the Electric Age." "We have absolutely disrupted our lives over and over again."[10]

McLuhan accurately foresaw that electronic media would be combined with computers to spread information cheaply and instantly around the world, in a variety of forms. Thirty years before the internet was launched, he wrote: "The next medium, whatever it is — it may be the extension of consciousness — will include television as its content, not as its environment, and will transform television into an art form. A computer as a research and communication instrument could enhance retrieval, obsolesce mass library organization, retrieve the individual's encyclopedic function and flip into a private line to speedily tailored data of a saleable kind."[11] Furthermore, McLuhan argued optimistically that we might regain the breadth of our senses which the printed word had diminished, restoring the preliterate "tribal balance" between all of our senses through a unified, "seamless web" of experience. As electronic communication connected us, the world would become a "global village" — another of McLuhan's catchphrases.

McLuhan owed a pronounced intellectual debt to a visionary and mystic who came before him: Teilhard de Chardin. A Jesuit priest, a geologist, and a paleontologist who played a role in the discovery of Peking man, de Chardin took a very big-picture view of the universe and our place within it, a picture that encompassed and motivated some of McLuhan's major insights. De Chardin also foresaw global communications and the internet, writing in the 1950s about "the extraordinary network of radio and television communication which already link us all in a sort of 'etherised' human consciousness," and "those astonishing electronic computers which enhance the speed of thought and pave the way for a revolution in the speed of research." This technology, he wrote, was creating a "nervous system for humanity," a "stupendous thinking machine." "The age of civilisation has ended," he said, "and the age of *one civilisation* is beginning."[12]

These ideas were an extension of de Chardin's magnum opus, *The Phenomenon of Man*. He completed the manuscript in the late 1930s, but because of his heterodox views, his ecclesiastical order refused throughout his lifetime to permit him to publish any of his writings. So de Chardin's books, and many collections of his essays, were only published after his death in 1955.

In spite of being a Catholic priest, de Chardin accepted Darwinian evolution as fact, and he built his futuristic vision around it. He saw the physical universe as in a state of constant evolution. Indeed, *The Phenomenon of Man* presents a "history of the universe" in terms that are surprisingly modern. De Chardin was probably influenced in this by another Jesuit priest, the founder of the hot big bang cosmology, Georges Lemaître.

De Chardin describes the emergence of complexity in the universe, from particles to atoms to molecules, to stars and planets, complex molecules, living cells, and consciousness, as a progressive "involution" of matter and energy, during which the universe becomes increasingly self-aware. Humans are self-aware and of fundamental significance to the whole. De Chardin quotes with approval Julian Huxley, who stated that "Man discovers that *he is nothing else than evolution become conscious of itself.*"[13] Huxley was the grandson of T. H. Huxley, the biologist famously known as "Darwin's bulldog" for his articulate defence of evolutionary theory in the nine-teenth century. He was also one of the founders of the "modern evolutionary synthesis," linking genetics to evolution. De Chardin took Huxley's statement to a cos-mic scale, envisioning that human society, confined to the Earth's spherical surface, would become increasingly connected into what would be in effect a very large liv-ing cell. With its self-consciousness and its inventions, it would continue to evolve through non-biological means towards an ultimate state of universal awareness, which he called the "Omega Point."

De Chardin's arguments are vague, allusive, and (despite his claims) necessarily unscientific, since many key steps, such as the formation of cells and life, and the emergence of consciousness, are well beyond our scien-tific understanding, as, of course, is the future. His vision is nonetheless interesting for the way in which it sees in evolution a latent potential for progress towards increas-ing complexity within the physical substance of the world. This potential is becoming increasingly evident as human advancement through technology and col-laboration supercedes survival of the biologically fittest as the driver of evolutionary progress. As Huxley says in his introduction to de Chardin's book, "We, mankind, contain the possibilities of the earth's immense future, and can realise more and more of them on condition that we increase our knowledge and our love. That, it seems to me, is the distillation of *The Phenomenon of Man*."[14]

McLuhan and de Chardin accurately foresaw the digital age and the future impact of electronic communication on the evolution of society. As McLuhan put it, "The medium, or process, of our time — electric technology — is reshaping and restructuring patterns of our social interdependence and every aspect of our personal life... Everything is changing — you, your family, your neighbourhood, your job, your government, your relation to 'the others.' And they're changing dramatically." He also foresaw some of the features and dangers of the internet and social media. He described an "electrically computerized dossier bank — that one big gossip column that is unforgiving, unforgetful, and from which there is no redemption, no erasure of early 'mistakes.'"[15]

These comments are insightful. They point to the clash between digital information and our analog nature. Our bodies and our senses work in smooth, continuous ways, and we most appreciate music or art or natural experiences that incorporate rich, continuous textures. We are analog beings living in a digital world, facing a quantum future.

DIGITAL INFORMATION IS THE crudest, bluntest, most brutal form of information that we know. Everything can be reduced to finite strings of 0s and 1s. It is completely unambiguous and is easily remembered. It reduces everything to black and white, yes or no, and it can be copied easily with complete accuracy. Obviously, analog information is infinitely richer. One analog number can take an infinite number of values, infinitely more values than can be taken by any finite number of digital bits.

The transition from analog to digital sound — from records and tapes to CDs and MP3s — caused a controversy, which continues to this day, about whether a digital reproduction is less rich and interesting to listen to than an analog version. By using more and more digital bits, one can mimic an analog sound to any desired accuracy. The fact remains that analog sound is inherently more subtle and less jarring than digital. Certainly, even in this digital age, analog instruments show no signs of going out of fashion.

Life's DNA code is digital. Its messages are written in three-letter "words" formed from a four-letter alphabet. Every word codes for an amino acid, and each sentence codes for a protein, made up of a long string of amino acids. The proteins form the basic machinery of life, part of which is dedicated to reading and transcribing DNA into yet more proteins. Although it is indeed amazing that all of the extravagant diversity and beauty of life is encoded in this way, it is also important to realize that the DNA code itself is not in any way alive.

Although the genetic basis for life is digital, living beings are analog creatures. We are made of plasmas, tissues, membranes, controlled by chemical reactions that depend continuously on concentrations of enzymes and reactants. Our DNA only comes to life when placed in an environment with the right molecules, fluids, and sources of energy and nutrients. None of these factors can be described as digital. New DNA sequences only arise as the result of mutations and reshufflings, which are partly environmental and partly quantum mechanical in origin. Two of the key processes that drive evolution — variation and selection — are therefore *not* digital. The main feature of the digital component of life — DNA — is its persistent, unambiguous character; it can be reproduced and translated into RNA and protein accurately and efficiently. The human body contains tens of trillions of cells, each with an identical copy of the DNA. Every time a cell divides, its DNA is copied.

It is tempting to see the digital DNA code as the fundamental basis of life, and our living bodies as merely its "servants," with our only function being to preserve our DNA and to enable its reproduction. But it seems to me that one can equally well argue that life, being fundamentally analog, uses digital memory simply to preserve the accuracy of its reproduction. That is, life is a happy combination of mainly digital memory and mainly analog operations.

At first sight, our nerves and brains might appear to be digital, since they either fire or do not in response to stimuli, just as the basic digital storage element is either 0 or 1. However, the nerve-firing rate can be varied continuously, and nerves can fire either in synchrony or in various patterns of disarray. The concentrations and flows of biomolecules involved in key steps, such as the passage of signals across synapses, are analog quantities. In general, our brains appear to be much more nuanced and complex systems than digital processors. This disjuncture between our own analog nature and that of our computers is quite plausibly what makes them so dissatisfying as companions.

Although analog information can always be accurately mimicked by using a sufficient number of digital bits, it is nevertheless a truism that analog information is infinitely richer than digital. Quantum information is infinitely richer again. Just one qubit of quantum information is described by a continuum of values. As we increase the number of qubits, the number of continuous values required to describe them grows exponentially. The state of a 300-qubit quantum computer (which might consist of a chain of just 300 atoms in a row) would be described by more numbers than we could represent in an analog manner, even if we used the three-dimensional position of every single one of the 10^{90} or so particles in the entire visible universe.

The ability of physical particles to carry quantum information has other startling consequences, stemming from entanglement, in which the quantum state of two particles is intrinsically interlinked. In Chapter Two, I described how, in an Einstein–Podolsky–Rosen experiment, two particles fly apart with their spins "entangled," so that if you observe both particles' spin along some particular axis in space, then you will always find one particle's spin pointing up while the other points down. This correlation, which Einstein referred to as "spooky action at a distance," is maintained no matter how far apart the particles fly. It is the basis for Bell's Theorem, also described in Chapter Two, which showed that the predictions of quantum theory can never be reproduced by classical ideas.

Starting in the 1980s, materials have been found in which electrons exhibit this strange entanglement property *en masse*. The German physicist Klaus von Klitzing discovered that if you suspend a piece of semiconductor in a strong magnetic field at a very low temperature, then the electrical conductance (a measure of how easily electric current flows through the material) is quantized. That is, it comes in whole number multiples of a fundamental unit. This is a very strange result, like turning on a tap and finding that water will flow out of it only at some fixed rate, or twice that rate, or three times the rate, however you adjust the tap. Conductance is a property of large things: wires and big chunks of matter. No one expected that it too could be quantized. The importance of von Klitzing's discovery was to show that in the right conditions, quantum effects can still be important, even for very large objects.

Two years later, the story took another twist. The German physicist Horst Störmer and the Chinese physicist Dan Tsui, working at Bell Labs, discovered that the conductance could also come in rational fractions of the basic unit of conductance, fractions like ⅓, ⅖, and 3/7.

The U.S. theorist Robert Laughlin, working at Stanford, interpreted the result as being due to the collective behaviour of all the electrons in the material. When they become entangled, they can form strange new entities whose electric charge is given in fractions of the charge of an electron.

Ever since these discoveries, solid state physicists have been discovering more and more examples of systems in which quantum particles behave in ways that would be classically impossible. These developments are challenging the traditional picture of *individual* particles, like electrons, carrying charge through the material. This picture guided the development of the transistor, but it is now seen as far too limited a conception of the possible states of matter. Quantum matter can take an infinitely greater variety of forms. The potential uses of these entirely new states of matter, which, as far as we know, never before formed in the universe, are only starting to be explored. They are likely to open a new era of quantum electronics and quantum devices, capable of doing things we have never seen before.

IN THE EARLY TWENTIETH century, the smallest piece of matter we knew of was the atomic nucleus. The largest was our galaxy. Over the subsequent century, our most powerful microscopes and telescopes have extended our view down to a ten-thousandth the size of an atomic nucleus and up to a hundred thousand times the size of our galaxy.

In the past decade, we have mapped the whole visible universe out to a distance of nearly fourteen billion light years. As we look farther out into space, we see the universe as it was longer and longer ago. The most distant images reveal the infant universe emerging from the big bang, a hundred-thousandth of its current age. It was extremely uniform and smooth, but the density of matter varied by around one part in a hundred thousand from place to place. The primordial density variations appear to take the same form as quantum fluctuations of fields like the electromagnetic field in the vacuum, amplified and stretched to astronomical scales. The density variations were the seeds of galaxies, stars, planets, and, ultimately, life itself, so the observations seem to be telling us that quantum effects were vital to the origin of everything we can now see. The Planck satellite, currently flying and due to announce its results soon, has the capacity to tell us whether the very early universe underwent a burst of exponential expansion. Over the coming decades, yet more powerful satellite observations may be able to tell whether there was a universe before the big bang.

Very recently, the Large Hadron Collider has allowed us to probe the structure of matter on the tiniest scales ever explored. In doing so, it has confirmed the famous Higgs mechanism, responsible for determining the properties of the different types of elementary particles. Beyond the Large Hadron Collider, the proposed International Linear Collider will probe the structure of matter much more accurately on these tiniest accessible scales, perhaps revealing yet another layer of organization, such as new symmetries connecting matter particles and forces.

With experiments like the Large Hadron Collider and the Planck satellite, we are reaching for the inner and outer limits of the universe. Equally significant, with studies of quantum matter on more everyday scales, we are revealing the organization of entangled levels of reality more subtle than anything so far seen. If history is any guide, these discoveries will, over time, spawn new technologies that will come to dominate our society.

Since the 1960s, the evolution of digital computers has been inexorable. Moore's law has allowed them to shrink and move progressively closer to our heads, from freezer-sized cabinets to desktops to laptops to smartphones held in our hands. Google has just announced Project Glass, a pair of spectacles incorporating a fully capable computer screen. With the screen right next to your eye, the power requirements are tiny and the system can be super-efficient. No doubt the trend will continue and computers will become a more and more integral part of our lives, our bodies, and our selves.

Having access to vast stores of digital information and processing power is changing our society and our nature. Our future evolution will depend less and less on our biological genes, and more and more on our abilities to interact with our computers. The future battle for survival will be to program or be programmed.

However, we are analog creatures based upon a digital code. Supplementing ourselves with more and more digital information is in this sense evolutionarily regressive. Digital information's strongest feature is that it can be copied cheaply and accurately and translated unambiguously. It represents a reduction of analog information, the "dead" blueprint or memory of life, rather than the alive analog element.

On the other hand, quantum information is infinitely deeper, more subtle and delicate than the analog information familiar to us. Interacting with it will represent a giant leap forward. As I have already explained, a single qubit represents more information than any number of digital bits; three hundred qubits represents more information than could be encoded classically using every particle in the universe. But the flip side is that quantum information is extremely fragile. The laws of quantum physics imply that it cannot be copied, a result known as the "no cloning" theorem. Unlike classical computers, quantum computers will not be able to replicate themselves. Without us, or at least some classical partner, they will not be able to evolve.

So it seems that a relationship between ourselves, as analog beings, and quantum computers may be of great mutual benefit, and it may represent the next leap forward for evolution and for life. We shall provide the definiteness and persistence, while quantum computers embody the more flighty, exploratory, and wide-ranging component. We will ask the questions, and the quantum computer will provide the answers. Just as our digital genes encode our analog operations, we, or our evolutionary successors, shall be the "operating system" of quantum life.

In the same way our DNA is surrounded by analog machinery bringing it to classical life, we will presumably become surrounded by quantum computers, making us even more alive. The best combinations of people and quantum computers will be the most successful, and will survive and propagate. With their vast information-processing capacities, quantum computers may be able to monitor, repair, or even renew our bodies. They will allow us to run smart systems to ensure that energy and natural resources are utilized with optimal efficiency. They will help us to design and oversee the production of new materials, like carbon fibres for space elevators and antimatter technologies for space propulsion. Quantum life would seem to have all the qualities needed to explore and understand the universe.

. . .

WHILE THE POSSIBILITY OF a coming "Quantum Age" is exciting, nothing is guaranteed about the future: it will be what we make of it. For a sharp dose of pessimism, let us turn to a remarkable woman visionary whose main targets were the Romantic notions of her age, the Victorian "Age of Wonder" and exploration, and the Industrial Revolution.

Mary Shelley was the daughter of one of the first feminists, Mary Wollstonecraft, a philosopher, educator, and the author, in 1792, of *A Vindication of the Rights of Women;* her father was William Godwin, a radical political philosopher. During childbirth, Wollstonecraft contracted a bacterial infection, and she died soon after. Throughout her life, Shelley continued to revere her mother. She was raised by her father, and when sixteen years old she became involved with Percy Bysshe Shelley, one of England's most famous Romantic poets. Percy was already married, and their relationship caused a great scandal. After his first wife committed suicide, Percy married Mary. They had four children (two before they were married), although only the last survived. The first was premature and died quickly. The second died of dysentery and the third of malaria, both during their parents' travels in Italy.

Mary started writing *Frankenstein; or, The Modern Prometheus* on one of these journeys to Italy with Percy, when she was only eighteen. It was published anonymously, although with a preface by Percy, when Mary was twenty-one. Now recognized as one of the earliest works of science fiction,[16] *Frankenstein* provides a compelling warning about the seductions and dangers of science. Shelley's reference to Prometheus shows the still-persistent influence of ancient Greek civilization on the most forward thinkers of the time.

Contemporary science set the background for the novel. In November 1806, the British chemist Sir Humphrey

Davy gave the Bakerian Lecture at the Royal Society in London. His topic was electricity and electrochemical analysis. In his introduction he said, "It will be seen that Volta [inventor of the battery] has presented to us a key that promises to lay open some of the most mysterious recesses of nature...There is now before us a boundless prospect of novelty in science; a country unexplored, but noble and fertile in aspect; a land of promise in philosophy."[17]

Public demonstrations and experiments were very popular in London at this time. A particularly notorious example was an attempt by another Italian, Giovanni Aldini, professor of anatomy at Bologna, to revive the body of a murderer six hours after he had been hanged. Aldini's demonstrations were breathlessly reported in the press: "On the first application of the electrical arcs, the jaw began to quiver, the adjoining muscles were horribly contorted, and the left eye actually opened... Vitality might have been fully restored, if many ulterior circumstances had not rendered this — *inappropriate*."[18]

Mary Shelley was likely inspired by events like these, and the general fascination with science, to write *Frankenstein*. Her novel captures the intensity and focus of a young scientist — Dr. Frankenstein — on the track of solving a great mystery: "After days and nights of incredible labour and fatigue, I succeeded in discovering the cause of generation and life; nay, more, I became myself capable of bestowing animation upon lifeless matter." Overwhelmed with the exhilaration of his finding, he states, "What had been the study and desire of the wisest men since the creation of the world was now within my grasp." His success encourages him to press on: "My imagination was too much exalted by my first success to permit me to doubt of my ability to give life to an animal as complex and wonderful as man."[19] Without any thought of the possible dangers, Frankenstein creates a monster whose need for companionship he cannot satisfy and who eventually exacts revenge by murdering Frankenstein's new bride.

In 1822, four years after *Frankenstein* appeared, Percy Shelley was drowned in a sailboat accident off Italy. Four years after that, Mary published her fourth novel, *The Last Man*, foretelling the end of humankind in 2100 as the result of a plague. The book is a damning critique of man's romantic notions of his own power to control his destiny. Echoing *Frankenstein*'s reference to Prometheus, *The Last Man* opens with the discovery of the cave of an ancient Greek oracle at Cumae in southern Italy. (In fact, the cave was actually discovered more than a century after the publication of Shelley's book.) The narrator recounts finding scattered piles of leaves on which the Sibyl, or prophetess, of the oracle at Cumae, recorded her detailed premonitions. After years of work organizing and deciphering the scattered fragments, the narrator presents *The Last Man* as a transcription of the Sibyl's predictions.

In her introduction, Shelley refers to Raphael's last painting, *The Transfiguration*. It depicts a stark dichotomy

between enlightenment and nobility in the upper half
of the painting, and the chaotic, dark world of humanity in the lower half. This conflict, between Apollonian
and Dionysian principles, has been one of the most constant themes in literature and in philosophy. Apollo and
Dionysus were both sons of Zeus. Apollo was the god
of the sun, dreams, and reason; Dionysus was the god
of wine and pleasure. Shelley's reference to the painting is interesting. There is a mosaic copy of Raphael's
Transfiguration in St. Peter's Basilica in Rome — a sort of
digital version of the real painting — and Shelley com-
pares her task of reconstructing the Sibyl's vision with
that of assembling *The Transfiguration* if all one had were
the painted tiles.

The central theme of the book is the failure of
Romantic idealism. Mary's husband, Percy, believed profoundly in the primacy of ideas. Writing about ancient
Rome, for example, he stated: "The imagination behold-
ing the beauty of this order, created it out of itself according to its own idea: the consequence was empire, and the
reward ever-living fame."[20]

Early on in *The Last Man*, the ambitious, fame-
seeking Raymond is elected as Lord Protector: "The new
elections were finished; parliament met, and Raymond
was occupied in a thousand beneficial schemes...he
was continually surrounded by projectors and projects
which were to render England one scene of fertility and
magnificence; the state of poverty was to be abolished;
men were to be transported from place to place almost

with the same facility as the Princes in the *Arabian Nights*..."[21]

Before any of these plans came to be realized, war between Greece and Turkey intervenes, and Raymond is killed in Constantinople. Adrian, son of the last king of England, is a leading figure. But he is a hopeless dreamer (clearly modelled after Shelley's husband, Percy). Following a brief period of peace, he states, "Let this last but twelve months... and earth will become a Paradise. The energies of man were before directed at the destruction of his species: they now aim at its liberation and preservation. Man cannot repose, and his restless aspirations will now bring forth good instead of evil. The favoured countries of the south will throw off the iron yoke of servitude [Shelley's reference to slavery]; poverty will quit us, and with that, sickness. What may not the forces, never before united, of liberty and peace achieve in this dwelling of man?"[22]

Adrian's dreams are also soon shattered. A plague is rapidly spreading west from Constantinople, and people come flooding in from Greece, Italy, and France. Raymond's dithering successor, Ryland, flees his position as the plague enters London. Adrian, the Romantic, assumes command, but his principal strategy is to convince people to pretend that there is no plague. Eventually the truth becomes obvious and he has no choice but to lead the population out of England, onto the Continent, where they slowly and painfully die.

Throughout, Shelley recounts the false optimism

of the characters, who are always trying to see a bright future when they are in fact doomed. Their grandiose visions and their delusions about the powers of reason, right, and progress cause them to fail, again and again. Finally, the last survivor, Verney, sails off around the world: "I form no expectation of alteration for the better, but the monotonous present is intolerable to me. Neither hope nor joy are my pilots — restless despair and fierce desire of change lead me on. I long to grapple with danger, to be excited by fear, to have some task, however slight or voluntary, for each day's fulfilment."[23] The irony is palpable. *The Last Man* was received poorly on its publication and was forgotten for one and a half centuries, but it has recently come to be seen as Shelley's second-most important work.

I cannot help also referring to one of the book's minor characters, the astronomer Merrival. His calculations have told him that in a hundred thousand years, the pole of the Earth will coincide with the pole of the Earth's orbit around the sun and, in his words, "a universal spring will be produced, and the earth will become a paradise."[24] He pays no attention to the spreading plague, even when it affects his family, busy as he is writing his "Essay on the Pericyclical Motions of the Earth's Axis." I hope we scientists are not Merrivals!

NEARLY TWO CENTURIES HAVE passed since *Frankenstein*. The monster Shelley envisaged Dr. Frankenstein creating has not materialized, nor so far have uncontrollable diseases like the plague she imagined in *The Last Man*. Nevertheless, the dangers she speaks about are as relevant as ever, and we would do well to pay attention to her concerns. Advances in biology have led to vaccinations, antibiotics, antiretrovirals, clean water, and other revolutionary public health advances. And genetic engineering has not yet produced any monsters. However, the benefits of science have been shared far too unevenly. There have been, and there continue to be, a vast number of deaths and untold suffering from preventable causes. The *only* guarantee of progress is a continued commitment to humane principles, and to conducting science on behalf of society.

One does not need to look far to find examples where science's success has encouraged a certain overreach and disconnect. There is a tendency to exaggerate the significance of scientific discoveries, and to dismiss nonscientific ideas as irrelevant.

As an example from my own field of cosmology, let me cite Lawrence Krauss's recent book, *A Universe from Nothing*. In it, he claims that recent observations showing that the universe has simple, flat geometry imply that it could have been created out of nothing. His argument is, in my view, based upon a technical gaffe, but that is not my point here. Through a misrepresentation of the physics, he leaps to the conclusion that a creator was not needed. The book includes an afterword by Richard Dawkins, hailing Krauss's argument as the final nail in the coffin for religion. Dawkins closes with, "If *On the*

Origin of Species was biology's deadliest blow to super-
naturalism [which is what Dawkins calls religion], we
may come to see *A Universe from Nothing* as the equivalent from cosmology. The title means exactly what it
says. And what it says is devastating."

The rhetoric is impressive, but the arguments
are shallow. The philosopher David Albert — one of
today's deepest thinkers on quantum theory — framed
his response at the right level, in his recent review of
Krauss's book in the *New York Times*, lamenting that "all
that gets offered to us now, by guys like these, in books
like this, is the pale, small, silly, nerdy accusation that
religion is, I don't know, *dumb*."[25] In comparing Krauss's
and Dawkins's arguments with the care and respect-
fulness of those presented by Hume in his *Dialogues
Concerning Natural Religion,* all the way back in the
eighteenth century, one can't help feeling the debate has
gone backwards. Hume presents his skepticism through
a dialogue which allows opposing views to be forcefully
expressed, but which humbly reaches no definitive conclusion. After all, that is his main point: we do not know
whether God exists. One of the participants is clearly
closest to representing Hume's own doubts: tellingly,
Hume names him Philo, meaning "love."

For another example of the disconnection between
science and society, let me quote the final paragraph
of U.S. theoretical physicist and Nobel Prize winner
Steven Weinberg's otherwise excellent book *The First
Three Minutes,* describing the hot big bang. He says,

"The more the universe seems comprehensible, the more it seems pointless. But if there is no solace in the fruits of our research, there is at least some consolation in the research itself...The effort to understand the universe is one of the very few things that lifts human life a little above the level of farce, and gives it some of the grace of tragedy."[26]

Many scientists express this viewpoint, that the universe seems pointless at a deep level, and that our situation is somehow tragic. For myself, I find this position hard to understand. Merely to be alive, to experience and to appreciate the wonder of the universe, and to be able to share it with others, is a miracle. I can only think that it is the separation of scientists from society, caused by the focus and intensity of their research, that leads them to be so dismissive of other aspects of human existence.

Of course, taking the view that the universe seems pointless is also a convenient way for scientists to eliminate, as far as possible, any prior prejudices or ulterior motives from their research. They want to figure out how things work without being biased by any thoughts of why they might work that way. It is reasonable to postpone questions of purpose when we have no scientific means of answering them. But to deny such influences is not to deal with them. Scientists are often consciously or unconsciously driven by agendas well outside science, even if they do not acknowledge them.

Many people outside science are interested in exactly the questions that scientists prefer to avoid. They want to know what scientific discoveries mean: in the case of cosmology, *why* the universe exists and *why* we are here. I think that if science is to overcome the disconnection with society, it needs to be better able to explain science's greatest lesson: that for the purpose of advancing our knowledge, it is extremely important to doubt constantly and to live with uncertainty. Richard Feynman put it this way: "This attitude of mind — this attitude of uncertainty — is vital to the scientist, and it is this attitude of mind which the student must first acquire. It becomes a habit of thought. Once acquired, we cannot retreat from it anymore."[27] In today's soundbite world, intellectual modesty and being frank about uncertainty are not the easiest things to promote. Nevertheless, I suspect scientists will become more, not less, credible if they do so, and society will feel less alienated from science.

My own view is that science should ultimately be about serving society's needs. Society needs to better understand science and to see its value beyond just providing the next gadget or technical solution. Science should be a part of fulfilling society's goals and creating the kind of world we would like to inhabit. Building the future is not only about servicing our needs, although those are important. There's an inspirational aspect of science and of understanding our place in the universe which enriches society and art and music and literature and everything else. Science, in its turn, becomes more creative and fruitful when it is challenged to explain what it is doing and why, and when scientists better appreciate the importance of their work to wider society.

Ever since the ancient Greeks, science has well appreciated that a free exchange of ideas, in which we are constantly trying out new theories while always remaining open to being proved wrong, is the best way to make progress. Within the scientific community, a new student can question the most senior professor, and authority is never acceptable as an argument. If our ideas are any good, it does not matter where they come from; they must stand on their own. Science is profoundly democratic in this sense. While its driver is often individual genius or insight, it engenders a strong sense of common cause and humility among its practitioners. These ways of thinking and behaving are valuable well beyond the borders of science.

However, as science has grown, it has also become increasingly specialized. To quote Richard Feynman again, "There are too few people who have such a deep understanding of two departments of our knowledge that they do not make fools of themselves in one or the other."[28] As science fragments, it becomes less accessible, both to other scientists and to the general public. Opportunities for cross-fertilization are missed, scientists lose their sense of wider purpose, and their science is reduced either to a self-serving academic exercise or a purely technical task, while society remains ignorant of science's great promise and importance.

There are ways of overcoming this problem of disconnection, and they are becoming increasingly important.

I AM FORTUNATE TO live in a very unusual community in Canada with a high level of public interest in science. Every month, our institute, Perimeter, holds a public lecture on physics in the local high school, in a hall with a capacity of 650. Month in and month out, the lectures are packed, with all the tickets sold out.

How did this happen? The key, I believe, is simply respect. When scientists make a serious attempt to explain what they are doing and why, it isn't hard to get people excited. There are many benefits: for the public, it is a chance to learn first-hand from experts about cutting-edge research; for scientists, it is a great chance to share one's ideas and to learn how to explain them to non-specialists. It is energizing to realize that people outside your field actually care. Finally, and most importantly, for young people, attending an exciting lecture can open the path to a future career.

In the heyday of Victorian science, many scientists engaged in public outreach. As we learned in Chapter One, Michael Faraday was recruited into science at a public lecture given by Sir Humphrey Davy at the Royal Institution in London. Faraday went on to succeed Davy as the director of the institution and give many public lectures himself. While a fellow at Cambridge, James Clerk Maxwell helped to found a workingmen's college providing scientific lectures in the evenings, and he persuaded local businesses to close early so their workers could attend. When he became a professor at Aberdeen and then King's College, London, he continued to give at least one evening lecture each week at the working-men's colleges there.[29]

Today, the internet provides an excellent medium for public outreach. One of the first students to attend the new master's program at our institute, Henry Reich, went on to pursue an interest in film. A year later, he launched a YouTube channel called MinutePhysics. It presents cleverly thoughtful, low-tech but catchy explanations of basic concepts in physics, making the ideas accessible and captivating to a wide audience. Henry realized there is a treasure trove of insights, many never before explained to the public, lying buried in the scientific literature. Communicating them well requires a great deal of care, thought, and respect for your audience. When quality materials are produced, people respond. Henry's channel now has more than three hundred thousand subscribers.

At our institute, we also engage in scientific inreach. The idea is to bring people from fields outside science, from history, art, music, or literature, into our scientific community. Science shares a purpose with these other disciplines: to explore and appreciate this universe we are privileged to inhabit. Every one of these human activities is inspiring, as they stretch our senses in different and complementary ways. However much any of us has learned, there is so much more that we do not know. What we have in common, in our motives and loves and aspirations, is much more important than any of our differences. Looking back on the great eras of discovery and progress, we see that this commonality of purpose was critical, and it seems to me we have to recreate it.

Throughout these chapters, we have looked at the special people, places, and times that produced profound progress. We have looked at ancient Greece, where a great flowering of science, philosophy, art, and literature went hand in hand with new ways of organizing society. The philosopher Epicurus, for example, seems in some respects to have anticipated the arguments of Hume and Galileo, arguing that nothing should be believed without being tested through direct observation and logical deduction; in other words, the scientific method.[30] Epicurus is also credited with the ethic of reciprocity, according to which one should treat others as one would like to be treated by them. These two ideas laid the foundations for justice: that everyone has the same right to be fairly treated and no one should be penalized until their crime is proven. Likewise, the methods and principles of scientific discourse were foundational to the creation of our modern democracy. We all have the capacity to reason, and everyone deserves an equal hearing.

We also looked at the Italian Renaissance, when the ancient Greek ideals were recovered and enlightenment progressed once more. In the Scottish Enlightenment, people encouraged each other to see the world confidently and with fresh eyes, to form new ways of understanding and representing it, and of teaching and communicating. These periods represented great liberations for society and great advances for science.

The enlightenments of the past did not begin in the most powerful countries: Greece was a tiny country, constantly threatened from the east and the north; Scotland was a modest neighbour of England. What they had in common was a sense among their people that this was their moment. They were countries that grasped an opportunity to become centres for reason and for progress. They had the courage to shape themselves and the future, and we are all still feeling the impact.

It is tempting to draw parallels between eighteenth-century Scotland, one step removed from its far more powerful colonial neighbour, and today's Canada, which, compared to the modern Rome to its south, feels like a haven of civilization. Canada has a great many advantages: strong public education and health care systems; a peaceful, tolerant, and diverse society; a stable economy; and phenomenal natural resources. It is internationally renowned as a friendly and peaceful nation, and widely appreciated for its collaborative spirit and for the modest, practical character of its people. There are many other countries and places in the world that hold similar promise, as centres to host the next great flowering of civilization on behalf of the planet. I can think of no better cause than for us to join together to make the twenty-first century unique as the era of the first Global Enlightenment.

· · ·

THE HISTORY OF PHYSICS traces back to the dawn of civilization. It is a story of how we have steadily realized our capacity to discover nature's deep secrets, and to build the understanding and the technologies that lay the basis for progress. Again and again, our efforts have revealed the fundamental beauty and simplicity in the universe. There is no sign of the growth in our knowledge slowing down, and what lies on the horizon today is every bit as thrilling as anything we have discovered in the past.

Today, we have many advantages over the scientists of earlier times. There are seven billion minds on the planet, mostly those of young people in aspiring, developing countries. The internet is connecting us all, providing instant access to educational and scientific resources. We need to be more creative in the ways we organize and promote science, and we need to allow more people to get involved. The world can become a hive of education, collaboration, and discussion. The entry of new cultures into the scientific community will be a vital source of energy and creativity.

We are better placed, too, to understand our position in the cosmos. We have just mapped the universe and pieced together the story of its emergence from a tiny ball of light some fourteen billion years ago. Likewise, we have detected the vacuum energy which dominates the universe and determines the Hubble length, the largest distance on which we will ever be able to see galaxies and stars. We have just discovered the Higgs particle, a manifestation of the detailed structure of the vacuum, predicted by theory half a century ago. Today, theory is poised to understand the big bang singularity and physics on the Planck length, a scale so tiny that classical notions of space and time break down.

All the indications are that the universe is at its simplest at the smallest and largest scales: the Planck length and the Hubble length. It may be no coincidence that the size of a living cell is the geometric mean of these two fundamental lengths. This is the scale of life, the realm we inhabit, and it is the scale of maximum complexity in the universe.

We live in a world with many causes of unhappiness. In these chapters, I have compared one of these, the information overload from the digital revolution, with the "ultraviolet catastrophe" that signalled classical physics' demise at the start of the twentieth century. One can draw further parallels with the selfish, individualistic behaviours that are often the root cause of our environmental and financial crises. Within physics, I see the idea of a "multiverse" as a similarly fragmented perspective, representing a loss of confidence in the prospects for basic science. Yet, I believe all of these crises will ultimately be helpful if they force us, like the quantum physicists, to remake our world in more holistic and far-sighted ways.

Through a deeper appreciation of the universe and our ability to comprehend it, not just scientists but everyone can gain. At a minimum, the magnificent cosmos provides some perspective on our parochial, human-created problems, be they social or political. Nature is organized in better ways, from which we can learn. The love of nature can bring us together and help us to appreciate that we are part of something far greater than ourselves. This sense of belonging, responsibility, and common cause brings with it humility, compassion, and wisdom. Society has too often been content to live off the fruits of science, without understanding it. Scientists have too often been happy to be left alone to do their science without thinking about why they are doing it. It is time to connect our science to our humanity, and in so doing to raise the sights of both. If we can only link our intelligence to our hearts, the doors are wide open to a brighter future, to a more unified planet with more unified science: to quantum technologies that extend our perception, to breakthroughs allowing us to access and utilize energy more cleverly, and to travel in space that opens new worlds.

What a privilege it is to be alive. Truly, we are faced with the opportunity of all time.

NOTES

CHAPTER ONE: MAGIC THAT WORKS

1. James Clerk Maxwell, quoted in Basil Mahon, *The Man Who Changed Everything: The Life of James Clerk Maxwell* (Chichester: Wiley, 2004), 48.

2. Aristotle, quoted in Kitty Ferguson, *Pythagoras: His Lives and the Legacy of a Rational Universe* (London: Icon Books, 2010), 108.

3. W. K. C. Guthrie, quoted in ibid., 74.

4. Richard P. Feynman, as told to Ralph Leighton, *"Surely You're Joking Mr. Feynman!": Adventures of a Curious Character*, ed. Edward Hutchings (New York: W. W. Norton, 1997), 132.

5. David Hume, *An Enquiry Concerning Human Understanding*, ed. Peter Millican (Oxford: Oxford University Press, 2008), 5.

6. Ibid., 6.

7. David Hume to "Jemmy" Birch, 1785, letter, quoted in E. C. Mossner, *The Life of David Hume* (Oxford: Oxford University Press, 2011), 626.

8. David Hume, *An Enquiry Concerning Human Understanding*, ed. Peter Millican (Oxford: Oxford University Press, 2008), 12.

9. Ibid., 45.

10. Ibid., 120.

11. Ibid., 45.

12. See, for example, "Geometry and Experience," Albert Einstein's address to the Prussian Academy of Sciences, Berlin, January 27, 1921, in *Sidelights on Relativity*., trans. G. B. Jeffery and W. Perrett (1922; repr., Mineola, NY: Dover, 1983), 8–16.

13. Leonardo da Vinci, *Selections from the Notebooks of Leonardo da Vinci*, ed. Irma Richter (London: Oxford University Press, 1971), 2.

14. Ibid., 7.

15. *The Notebooks of Leonardo da Vinci*, vol. 1, Wikisource, accessed July 4, 2012, http://en.wikisource.org/wiki/The_Notebooks_ of_Leonardo_Da_Vinci/I.

16. Albert Einstein, "Geometry and Experience" (address to the Prussian Academy of Sciences, Berlin, January 27, 1921), in *Sidelights on Relativity*, trans. G. B Jeffery and W. Perrett (1922; repr., Mineola, NY: Dover, 1983), 8.

17. Wikipedia, s.v. "Mathematical Beauty," accessed July 3, 2012, http://en.wikipedia.org/wiki/Mathematical_beauty.

18. For an interesting discussion of this, see Eugene P. Wigner, "The Unreasonable Effectiveness of Mathematics in the Natural Sciences" (Richard Courant Lecture in Mathematical Sciences, New York University, May 11, 1959), *Communications on Pure and Applied Mathematics* 13, no. 1 (1960):1–14.

19. Albert Einstein, quoted in Dava Sobel, *Galileo's Daughter: A Historical Memoir of Science, Faith, and Love* (New York: Walker, 1999), 326.

20. John Maynard Keynes, "Newton the Man," speech prepared for the Royal Society, 1946. See http://www-history.mcs.st-and. ac.uk/Extras/Keynes_Newton.html.

21. Arthur Herman, *How the Scots Invented the Modern World: The True Story of How Western Europe's Poorest Nation Created Our World and Everything in It* (New York: Three Rivers, 2001), 190.

22. George Elder Davie, *The Democratic Intellect: Scotland and Her Universities in the Nineteenth Century* (Edinburgh: Edinburgh University Press, 1964), 150.

23. J. Forbes, quoted in P. Harman, ed., *The Scientific Letters and Papers of James Clerk Maxwell*, vol. 1, *1846–1862* (Cambridge: Cambridge University Press, 1990), 8.

24. Alan Hirshfeld, *The Electric Life of Michael Faraday* (New York: Walker, 2006), 185.

25. John Meurig Thomas, "The Genius of Michael Faraday," lecture given at the University of Waterloo, 27 March 2012.

26. These are Cartesian coordinates, invented by the French philosopher René Descartes.

27. Michael Faraday to J. C. Maxwell, letter, 25 March 1857, in P. Harman, ed., *The Scientific Letters and Papers of James Clerk Maxwell*, vol. 1, *1846–1862* (Cambridge: Cambridge University Press, 1990), 548.

28. Alan Hirshfeld, *The Electric Life of Michael Faraday* (New York: Walker, 2006), 185.

29. J. C. Maxwell to Michael Faraday, letter, 19 October 1861, in P. Harman, ed., *The Scientific Letters and Papers of James Clerk Maxwell*, vol. 1, *1846–1862* (Cambridge: Cambridge University Press, 1990), 684–86.

CHAPTER TWO: OUR IMAGINARY REALITY

1. John Bell, "Introduction to the Hidden-Variable Question" (1971), in *Quantum Mechanics, High Energy Physics and Accelerators: Selected Papers of John S. Bell (with Commentary)*, ed. M. Bell, K. Gottfried, and M. Veltman (Singapore: World Scientific, 1995), 716.

2. Albert Einstein, "How I Created the Theory of Relativity," trans. Yoshimasa A. Ono, *Physics Today* 35, no. 8 (1982): 45–7.

3. Carlo Rovelli, *The First Scientist: Anaximander and His Legacy* (Yardley, PA: Westholme, 2011).

4. Wikipedia, s.v. "Anaximander," accessed April 15, 2012, http://en.wikipedia.org/wiki/Anaximander, and "Suda," accessed April 15, 2012, http://en.wikipedia.org/wiki/Suda.

5. Werner Heisenberg, "Quantum-Mechanical Re-interpretation of Kinematic and Mechanical Relations," in *Sources of Quantum Mechanics*, ed. B. L. van der Waerden (Amsterdam: North-Holland, 1967), 261–76.

6. Werner Heisenberg, quoted in J. C. Taylor, *Hidden Unity in Nature's Laws* (Cambridge: Cambridge University Press, 2001), 225.

7. Lauren Redniss, *Radioactive: Marie and Pierre Curie: A Tale of Love and Fallout* (HarperCollins, 2010), 17.

8. Werner Heisenberg, quoted in F. Selleri, *Quantum Paradoxes and Physical Reality* (Dordrecht, Netherlands: Kluwer, 1990), 21.

9. Wikipedia, s.v., "Max Planck," accessed July 10, 2012, http:// en.wikipedia.org/wiki/Max_Planck.

10. Ibid.

11. Ibid.

12. Albert Einstein, quoted in Abraham Pais, *Inward Bound: Of Matter and Forces in the Physical World* (New York: Oxford University Press, 1986), 134.

13. Susan K. Lewis and Neil de Grasse Tyson, "Picturing Atoms" (transcript from *NOVA ScienceNOW*), PBS, accessed July 4, 2012, http://www.pbs.org/wgbh/nova/physics/atoms-electrons.html.

14. Clifford Pickover, *The Math Book: From Pythagoras to the 57th Dimension, 250 Milestones in the History of Mathematics* (New York: Sterling, 2009), 118–24.

15. Richard P. Feynman, Robert B. Leighton, and Matthew Sands, *The Feynman Lectures on Physics*, vol. 1 (Reading, MA: Addison-Wesley, 1964), 22.

16. Werner Heisenberg, "Ueber den anschaulichen Inhalt der quantentheoretischen Kinematik and Mechanik," *Zeitschrift für Physik* 43 (1927), 172–98. English translation in John Archibald Wheeler and Wojciech H. Zurek, eds., *Quantum Theory and Measurement* (Princeton, NJ: Princeton University Press, 1983), 62–84.

17. There is a beautiful animation of diffraction and interference from two slits on Wikipedia, s.v. "Diffraction," accessed July 2, 2012, http://en.wikipedia.org/wiki/Diffraction.

18. F. Scott Fitzgerald, *The Crack-Up* (New York: New Directions, 1993), 69.

19. Werner Heisenberg, *Physics and Beyond: Encounters and Conversations* (New York: Harper & Row, 1971), 81.

20. Irene Born, trans., *The Born-Einstein Letters, 1916–1955: Friendship, Politics and Physics in Uncertain Times* (New York: Walker, 1971), 223.

21. My discussion here is a simplified version of David Mermin's simplified version of Bell's Theorem, presented in N. D. Mermin, "Bringing Home the Atomic World: Quantum Mysteries for Anybody," *American Journal of Physics* 49, no. 10 (1981): 940. See also Gary Felder, "Spooky Action at a Distance " (1999), North Carolina University, accessed July 4, 2012, http://www4. ncsu.edu/unity/lockers/users/f/felder/public/kenny/papers/ bell.html.

22. H. Minkowski, "Space and Time," in H. A. Lorentz, A. Einstein, H. Minkowski, and H. Weyl, *The Principle of Relativity*, trans. W. Perrett and G. B. Jeffery (1923; repr., Mineola, NY: Dover Publications, 1952), 75–91.

CHAPTER THREE: WHAT BANGED?

1. Thomas Huxley, "On the Reception of the Origin of Species" (1887), in Francis Darwin, ed., *The Life and Letters of Charles Darwin*, vol. 1 (New York: Appleton, 1904), accessed online at http://www.gutenberg.org/files/2089/2089-h/2089-h.htm.

2. John Archibald Wheeler, "How Come the Quantum?" *Annals of the New York Academy of Sciences* 480, no. 1 (1986): 304–16.

3.   Albert Einstein, quoted in Antonina Vallentin, *Einstein: A Biography* (Weidenfeld & Nicolson, 1954), 24.

4.   Luc Ferry, *A Brief History of Thought: A Philosophical Guide to Living* (New York: Harper Perennial, 2011), 19.

5.   Albert Einstein, "Über einen die Erzeugung und Verwandlung des Lichtes betreffenden heuristischen Gesichtspunk," *Annalen der Physik* 17, no. 6 (1905), 132–48. A good Wikisource translation is available online at http://en.wikisource.org/wiki/On_a_Heuristic_Point_of_View_about_the_Creation_and_Conversion_of_Light.

6.   Albert Einstein, "Maxwell's Influence on the Development of the Conception of Physical Reality," in *James Clerk Maxwell: A Commemorative Volume* (New York: Macmillan, 1931), 71.

7.   Max Planck invented so-called Planck units when thinking of how to combine gravity with quantum theory. The Planck scale is $L_p = (hG/c^3)^{1/2} = 4 \times 10^{-35}$ metres, a combination of Newton's gravitational constant; Planck's constant, h; and the speed of light, c. Below the Planck length, the effects of quantum fluctuations become so large that any classical notion of space and time becomes meaningless. The Planck energy is the energy associated with a quantum of radiation with a wavelength equal to the Planck length, $E_p = (hc^5/G)^{1/2} = 1.4$ MWh.

8.   Albert Einstein, quoted in Frederick Seitz, "James Clerk Maxwell (1831–1879), Member APS 1875," *Proceedings of the American Philosophical Society* 145, no. 1 (2001): 35. Available online at: http://www.amphilsoc.org/sites/default/files/Seitz.pdf.

9.   Albert Einstein and Leopold Infeld, *The Evolution of Physics* (New York: Simon & Schuster, 1938), 197–8.

10. John Archibald Wheeler and Kenneth William Ford, *Geons, Black Holes, and Quantum Foam: A Life in Physics* (New York: W. W. Norton, 2000), 235.

11. George Bernard Shaw, "You Have Broken Newton's Back," in *The Book of the Cosmos: Imagining the Universe from Heraclitus to Hawking*, ed. D. R. Danielson (New York: Perseus, 2000), 392–3.

12. Irene Born, trans., *The Born-Einstein Letters, 1916–1955: Friendship, Politics and Physics in Uncertain Times* (New York: Walker, 1971), 223.

13. John Farrell, *The Day Without Yesterday: Lemaître, Einstein, and the Birth of Modern Cosmology* (New York: Basic Books, 2010), 10.

14. Ibid, 207.

15. Abbé G. Lemaître, "Contributions to a British Association Discussion on the Evolution of the Universe," *Nature* 128 (October 24, 1931), 704–6.

16. Duncan Aikman, "Lemaître Follows Two Paths to Truth," *New York Times Magazine*, February 19, 1933.

17. Gino Segrè, *Ordinary Geniuses: Max Delbrück, George Gamow, and the Origins of Genomics and Big Bang Cosmology* (London: Viking, 2011), 146.

18. U.S. Space Objects Registry, accessed July 4, 2012, http:// usspaceobjectsregistry.state.gov/registry/dsp_DetailView.cfm.

19. Adam Frank, *About Time: Cosmology and Culture at the Twilight of the Big Bang* (New York: Free Press, 2011), 196–201.

20. Technically, this means that the cosmological constant is the unique type of matter that is Lorentz-invariant.

21. See Paul J. Steinhardt and Neil Turok, *Endless Universe: Beyond the Big Bang* (London: Weidenfeld & Nicolson, 2007).

22. Cicero, *On the Nature of the Gods*, Book II, Chapter 46, quoted in ibid., 171.

23. G. Lemaître, "L'Univers en expansion," *Annales de la Société Scientifique de Bruxelles* A21 (1933): 51.

CHAPTER FOUR: THE WORLD IN AN EQUATION

1. Paul Dirac, quoted in Graham Farmelo, *The Strangest Man: The Hidden Life of Paul Dirac, Mystic of the Atom* (New York: Basic Books, 2010), 435.

2. H. Weyl, "Emmy Noether," *Scripta Mathematica* 3 (1935): 201–20, quoted in Peter Roquette, "Emmy Noether and Hermann Weyl" (2008), an extended manuscript of a talk given at the Hermann Weyl Conference, Bielefield, Germany, September 10, 2006 (see http://www.rzuser.uni-heidelberg.de/~ci3/weyl+noether.pdf), 22.

3. Albert Einstein, "The Late Emmy Noether," letter to the editor of the *New York Times*, published May 4, 1935.

4. Helge Kragh, "Paul Dirac: The Purest Soul in an Atomic Age," in Kevin C. Knox and Richard Noakes, eds., *From Newton to Hawking: A History of Cambridge University's Lucasian Professors of Mathematics* (Cambridge: Cambridge University Press, 2003), 387.

5. John Wheeler, quoted by Sir Michael Berry in an obituary of Dirac. Available online at http://www.phy.bris.ac.uk/people/berry_mv/the_papers/Berry130.pdf .

6. P. A. M. Dirac, "The Evolution of the Physicist's Picture of Nature," *Scientific American* 208, no. 5 (May 1963): 45–53.

CHAPTER FIVE: THE OPPORTUNITY OF ALL TIME

1. The sole surviving fragment of Anaximander's works, as quoted by Simplicius (see http://www.iep.utm.edu/anaximan/#H4).

2. Louis C. K., during an appearance on *Late Night with Conan O'Brien,* originally aired on NBC on February 24, 2009.

3. John Gertner, *The Idea Factory: Bell Labs and the Great Age of American Innovation* (New York: Penguin, 2012).

4. Ibid., 149–52.

5. See the 1956 Nobel Prize lectures by Shockley, Brattain, and Bardeen, all of which are available online at http://www.nobelprize.org/nobel_prizes/physics/laureates/1956/.

6. Michael Riordan and Lillian Hoddeson, *Crystal Fire: The Invention of the Transistor and the Birth of the Information Age* (New York: W. W. Norton, 1997), 115–41.

7. Sebastian Loth et al., "Bistability in Atomic-Scale Antiferromagnets," *Science* 335, no. 6065 (January 2012): 196. For a lay summary, see http://www.ibm.com/smarterplanet/us/en/smarter_computing/article/atomic_scale_memory.html.

8. In fact, the quantum state of a qubit is specified by two real numbers, giving the location on a two-dimensional sphere.

9. A theorem due to Euclid, called the fundamental theorem of arithmetic, shows that such a factoring is unique.

10. Marshall McLuhan, *Understanding Me: Lectures and Interviews*, ed. Stephanie McLuhan and David Staines (Toronto: McClelland & Stewart, 2005), 56.

11. Marshall McLuhan and Bruce Powers, *Global Village: Transformations in World Life and Media in the 21st Century* (New York: Oxford University Press, 1992), 143.

12. Pierre Teilhard de Chardin, quoted in Tom Wolfe's foreword to Marshall McLuhan, *Understanding Me: Lectures and Interviews*, ed. Stephanie McLuhan and David Staines (Toronto: McClelland & Stewart, 2005), xvii.

13. Pierre Teilhard de Chardin, *The Phenomenon of Man* (Harper Colophon, 1975), 221.

14. Julian Huxley, in introduction to Pierre Teilhard de Chardin, *The Phenomenon of Man* (HarperCollins Canada, 1975), 28.

15. Marshall McLuhan and Quentin Fiore, *The Medium Is the Massage: An Inventory of Effects* (Toronto: Penguin Canada, 2003), 12.

16. Brian Aldiss, *The Detached Retina: Aspects of SF and Fantasy* (Liverpool: Liverpool University Press, 1995), 78.

17. Richard Holmes, *The Age of Wonder: How the Romantic Generation Discovered the Beauty and Terror of Science* (London: HarperPress, 2008), 295.

18. Ibid., 317.

19. Mary Shelley, *Frankenstein*, 3rd ed. (1831; repr., Mineola, NY: Dover, 1994), 31–2.

20. Percy Bysshe Shelley, "A Defence of Poetry" (1821), available online at http://www.bartleby.com/27/23.html.

21. Mary Shelley, *The Last Man* (1826; repr., Oxford: Oxford University Press, 2008), 106.

22. Ibid., 219.

23. Ibid., 470.

24. Ibid., 220.

25. D. Albert, "On the Origin of Everything," *New York Times*, March 23, 2012.

26. Steven Weinberg, *The First Three Minutes: A Modern View of the Origin of the Universe* (New York: Basic Books, 1977), 144.

27. Richard Feynman, *The Pleasure of Finding Things Out* (London: Penguin, 2007), 248.

28. Richard Feynman, "The Uncertainty of Science," in *The Meaning of It All: Thoughts of a Citizen Scientist* (New York: Perseus, 1998), 3.

29. Basil Mahon, *The Man Who Changed Everything: The Life of James Clerk Maxwell,* (Chichester: Wiley, 2004), 45.

30. See, for example, Elizabeth Asmis, *Epicurus' Scientific Method* (Ithaca, NY: Cornell University Press, 1984).

FURTHER READING

Albert, David. *Quantum Mechanics and Experience*. Cambridge, MA: Harvard University Press, 1994.

Deutsch, David. *The Beginning of Infinity*. New York: Viking, 2011.

Diamandis, Peter H., and Steven Kotler. *Abundance: The Future is Better Than You Think*. New York: Free Press, 2012.

Falk, Dan. *In Search of Time: Journeys Along a Curious Dimension*. Toronto: McClelland & Stewart, 2008.

Gowers, Timothy. *Mathematics*. New York: Sterling, 2010.

Greene, Brian. *The Fabric of the Cosmos: Space, Time and the Texture of Reality*. New York: Vintage, 2005.

Guth, Alan. *The Inflationary Universe: The Quest for a New Theory of Cosmic Origins*. New York: Basic Books, 1998.

Hawking, Stephen. *A Brief History of Time*. New York: Bantam, 1998.

Penrose, Roger. *The Road to Reality: A Complete Guide to the Laws of the Universe*. New York: Vintage, 2007.

Sagan, Carl. *Cosmos*. New York: Ballantine, 1985.

Steinhardt, Paul J., and Neil Turok. *Endless Universe: Beyond the Big Bang — Rewriting Cosmic History*. New York: Broadway, 2008.

Weinberg, Steven. *The First Three Minutes: A Modern View of the Origin of the Universe*. New York: Basic Books, 1993.

Zeilinger, Anton. *Dance of the Photons: From Einstein to Quantum Teleportation*. New York: Farrar, Straus & Giroux, 2010.

PERMISSIONS

Permission is gratefully acknowledged to reprint the following images:

Glenlair © Courtesy of Cavendish Laboratory, Cambridge

Force Field © Neil Turok

Raphael (Raffaello Sanzio) (1483–1520). *The School of Athens.* ca. 1510–1512. Fresco. © Scala/Art Resource, NY

Fifth Solvay Conference 1927 © Photograph by Benjamin Couprie, Institut International de Physique Solvay, courtesy AIP Emilio Segre Visual Archives

Double-Slit Experiment © Neil Turok

Big bang © Neil Turok

COBE Temperature Fluctuations © NASA

Dark Matter © NASA

AIMS South Africa © African Institute for Mathematical Sciences

ACKNOWLEDGEMENTS

I WOULD LIKE TO express my heartfelt thanks to my friends and colleagues at the Perimeter Institute for Theoretical Physics, a place for quantum leaps in space and time. Their constant encouragement and steadfast support kept me going as I struggled to prepare this manuscript. Once again, they made me realize how fortunate I am to be a part of this unique community. A special thanks is due to Mike Lazaridis, Perimeter's founder and the most visionary supporter our field ever had, and to those who ensure the institute maintains the highest standards of management and communications, including Michael Duschenes and John Matlock.

Throughout this project, Alexandra Castell lent me continuous assistance. Natasha Waxman played a major role researching and helping to prepare early drafts, ably assisted by Erin Bow and Ross Diener. Daniel Gottesman, Lucien Hardy, Adrian Kent, Rob Myers, Lee Smolin, and Paul Steinhardt generously read drafts and provided invaluable comments. I have benefitted from discussions with many scientific colleagues on these topics, including Itzhak Bars, Laurent Freidel, Stephen Hawking, Ray Laflamme, Sandu Popescu, and Xiao-Gang Wen. Malcolm Longair very kindly shared with me the proofs of his fascinating new book on the historical origins of quantum mechanics, *Quantum Concepts in Physics*. Thank you for your enthusiasm and your wisdom. Naturally, whatever errors and misconceptions remain in this book are entirely my own. Many thanks to Chris Fach and Erick Schnetter for help preparing the illustrations.

A huge thank you to all my partners in the African Institute for Mathematical Sciences (AIMS) project, and to all our wonderful students. Let me mention in particular Barry Green and Thierry Zomahoun. It is a constant pleasure to work with you and for you. I thank you for your patience and understanding during the writing of this book, and for your tireless commitment to our shared cause.

Philip Coulter at the CBC and Janie Yoon at House of Anansi Press deserve special gratitude for stepping in with inspirational advice at a critical time. Janie in particular found the right combination of praise and tough love to keep me on track. If this manuscript is at all readable, it is due to your heroic efforts.

And last but first, big hugs to Corinne and Ruby without whom I would be lost.

INDEX